Heat Engines and Applied Heat

for

National Certificate Students

Heat Engines
and Applied Heat
for
National Certificate Students

S I metric edition

F. METCALFE
B.Sc.(Eng.) Hons., C.Eng., M.I.Mech.E.

Director, Engineering Industry
Training Board
formerly: Principal, Ipswich Civic College

CASSELL · LONDON

Cassell Ltd
35 Red Lion Square, London WC1R 4SG
and at Sydney, Auckland, Toronto, Johannesburg

an affiliate of Macmillan Publishing Co. Inc., New York

First published as Heat Engines, 1960
Revised and extended and first published as Heat Engines and
    Applied Heat, 1966
Fourth revised edition, 1969
Fourth edition, second impression, January 1971
Fifth edition, June 1972
Fifth edition, second impression, July 1973
Fifth edition, third impression, December 1974
Fifth edition, fourth impression, May 1977
Fifth edition, fifth impression, February 1979

I.S.B.N. 0 304 29081 5

Printed and bound in Great Britain at
The Camelot Press Ltd, Southampton

PREFACE TO THE THIRD EDITION

Much of this book previously appeared as *Heat Engines* by the same author. It was decided to carry out a complete revision so as to take account of the so-called 'new' approach in thermodynamics in which the concept of heat and work as energy in transition is used. Many Ordinary National Certificate students will study the subject further in Degree and Higher National courses, and it seems sensible to use an approach which will be acceptable later on. In addition, the scope of the work has been enlarged to include the additional requirements of syllabuses for the O 2 stage of the subjects of Applied Heat and Heat Engines for Ordinary National Certificates and Diplomas in Engineering. Finally, the revision examples have been brought up to date with new material, and it is hoped that this new book will have the same encouraging reception as its predecessor.

1966   F. Metcalfe

PREFACE TO THE FOURTH EDITION

The Joint Committee for Ordinary National Certificates and Diplomas in Engineering has decided that from the 1970–71 Session all assessed examinations shall be set in S I units and in view of this, the examples and the relevant parts of Mr. Metcalfe's work have been revised into this system. Basically the text remains the same as in the last edition and the Revisor hopes the book will be held in the same regard as the work it replaces.

Thanks are due to the Institution of Mechanical Engineers, the Union of Educational Institutions and the Union of Lancashire and Cheshire Institutes for permission to modify past examination questions by conversion to S I units.

Taunton Technical College, 1969   F. R. Collinson

PREFACE TO THE FIFTH EDITION

The production of a fifth edition has given the opportunity to make some improvements in the text by eliminating some parts thought to be out of date and by adding new material. In particular the section on reciprocating steam engine design has given way to further work on the impulse-reaction steam turbine, and the section on radiation heat transfer has been extended.

E.I.T.B., 1972   F. Metcalfe

CONTENTS

CHAPTER   PAGE

Preface v

The S I System xi

1. Units and Definitions 1

   Revision Exercises 21

2. The Gas Laws 22

   Revision Exercises 33

3. Internal-combustion Engines 35

   Revision Exercises 64

4. The Expansion and Compression of Gases . . . 72

   Revision Exercises 90

5. Cycles of Operation 98

   Revision Exercises 118

6. Fuels and Combustion 128

   Revision Exercises 157

7. The Properties of Steam 166

   Revision Exercises 185

8. Steam Plant 193

   Revision Exercises 208

9. The Steam Engine and the Steam Turbine . . . 215

   Revision Exercises 237

10. Introduction to Heat Transmission 243

    Revision Exercises 254

    Index 257

PRINCIPAL SYMBOLS USED

C heat capacity
c specific heat capacity constant
H enthalpy
h specific enthalpy coefficient of heat transfer
k coefficient of thermal conductivity
L length of stroke
l specific latent heat
M molecular weight
m mass
$\dot{m}$ mass flow
N rev/s
n firing strokes/s
p pressure
p_g gauge pressure
p_{at} atmospheric pressure
P_m mean effective pressure
Q quantity of heat
$\dot{Q}$ heat flow rate, i.e. quantity of heat per unit time
R characteristic gas constant cut-off ratio (steam engine)
r volume compression ratio
T absolute temperature
t empirical temperature
U internal energy
u specific internal energy peripheral speed of rotor blades
V volume
v specific volume velocity
W work
w weight
X relative velocity
x dryness fraction vertical distance from datum

α angle of inclination of nozzles
γ ratio of specific heats $\dfrac{c_p}{c_v}$
ϵ total emissivity
η efficiency
θ empirical temperature rotor blade angle
λ wavelength
ρ density
σ Stefan-Boltzmann constant
ω angular velocity

Abbreviations for words other than names of units

atmospheric	atm.
direct current	d.c.
kilogramme-molecule	kmol
brake power	b.p.
indicated power	i.p.
standard temperature and pressure	s.t.p.
absolute	abs.

Subscripts

f   refers to a property of a saturated liquid
g   refers to a property of a saturated vapour
fg   refers to the change of phase at constant pressure from saturated liquid to saturated vapour
sup   refers to a property of a superheated vapour

SYSTÈME INTERNATIONAL D'UNITÉS (INTERNATIONAL SYSTEM OF UNITS OR S I)

The S I system is based on six *basic* units:

Quantity	*Unit*	*Symbol*
Length	Metre	m
Mass	Kilogramme	kg
Time	Second	s
Electric current	Ampere	A
Temperature	Degree kelvin	K
Luminous intensity	Candela	cd

The *derived* units are such as to provide a coherent relationship one with the other:

The NEWTON is the unit of force and is that force which will accelerate 1 kilogramme mass at 1 metre per second per second

i.e. $$1 \text{ N} = 1 \text{ kg} \times 1 \text{ m/s}^2 = 1 \ \frac{\text{kgm}}{\text{s}^2}$$

(1 kilogramme mass falling freely will accelerate at 9·81 metres per second per second and therefore the force due to the weight of 1 kilogramme mass is 9·81 newtons.)

The unit of force enables us to define the unit of work. Since work is force times the distance moved in the line of action of the force, then 1 newton × 1 metre (called a JOULE) is the unit of work

i.e. $$1 \text{ J} = 1 \text{ N} \times 1 \text{ m}$$

The first law of thermodynamics may be written: 'The work done on a system plus the increase in energy in the system is equal to the heat supplied to the system'. It therefore follows that the units that are appropriate for work are also appropriate for energy and also for heat.

Power is defined as the rate of doing work and in the S I system a joule per second is named the WATT,

i.e. $$1 \text{ W} = 1 \text{ J/s} = 1 \text{ Nm/s}$$

From the basic unit of electric current, the ampere, and the derived unit of power, the watt, the definition of electric potential can be derived; the VOLT is that electric potential between two points of a conducting wire carrying a constant current of 1 ampere when the power dissipated between these points is equal to 1 watt,

i.e. $$1 \text{ W} = 1 \text{ A} \times 1 \text{ V}$$

It will be seen from the above relationships that in all cases the 'numbers' associated with each quantity are unity, that is, the relationships are *coherent*.

The names of the multiples and sub-multiples of the basic and derived units are related to the basic and derived units by the addition of prefixes which are the same irrespective of the basic units to which they are applied:

i.e.
    1 *kilo*metre = 1000 metres
    1 *kilo*newton = 1000 newtons
    1 *kilo*watt   = 1000 watts

xi

There are three recommendations which have been made which it is hoped will be universally adopted:

1. Multiples and submultiples of basic unit limited to 10^3 intervals.
2. Double prefixes to be avoided.
3. Denominators in compound units should be in basic or derived units. (Meganewtons per square metre is preferred to newtons per square millimetre.)

Factor by which the unit is multiplied	Prefix	Symbol
One thousand million $= 10^9$	giga	G
One million $= 10^6$	mega	M
$1000 = 10^3$	kilo	k
$0 \cdot 001 = 10^{-3}$	milli	m
$0 \cdot 000\ 001$ (one millionth) $= 10^{-6}$	micro	μ
One thousand-millionth $= 10^{-9}$	nano	n

Note: It is unfortunate that no new name has yet been given to the kilogramme. To adhere rigidly to the system we should have, for 1000 times unit mass, 1 kilo-kilogramme. The term megagramme is preferred.

1. Units and Definitions

The work of an engineer is limited unless he has a source of power to drive his machines or tools. This book deals with engines capable of producing power from the energy stored in fuel. However, before such a study can begin, it is necessary to be sure about a number of definitions and units which are essential for a proper understanding of the subject. We are familiar with most of these items in everyday life, but science demands that we be exact in our understanding if real progress is to be made.

ENERGY

Let us begin by seeing what is meant by 'Energy'. When we say that a person is full of energy we mean that he is ready to go into action, to work or to run and so on. The scientist says that energy can exist in a number of forms. For example, water stored at a high level possesses energy which can be used to drive a water turbine—this type is called POTENTIAL ENERGY. Again, anything which is moving possesses a type of energy called KINETIC ENERGY, the energy of motion (a person who leaves his thumb under a moving hammer is soon convinced of this!). It is the kinetic energy possessed by a high-speed jet of gases which gives the thrust to a jet aeroplane.

We shall need to understand that form of energy known as INTERNAL ENERGY which all bodies possess. It is known that all substances consist of a large number of independent units called molecules, and these molecules attract each other to form a mass. Each molecule is moving in a particular path so that it possesses kinetic energy, and the sum of such energy of all the molecules in a body is known as the internal energy of the body. The higher the temperature of a body, the faster its molecules are moving and correspondingly, the more internal energy it possesses. When a solid body is continuously heated, the speed of the molecules eventually

1

reaches a point at which they break away from their normal path so that the body melts and becomes a liquid. If the heating is continued further, the increased motion of the molecules in the liquid state reaches a point at which they leave the surface of the liquid, and the substance becomes a gas.

Finally we shall be concerned with two further forms of energy, HEAT ENERGY and WORK, which are the two particular forms that energy takes when it is being transferred from one system to another. Heat is the form in which energy flows from one system to another when there is a temperature difference between them. Work is done when a force moves through any distance and is the form in which energy moves from one system to another using a mechanical device.

SYSTEMS

We shall need to study *systems*, which we may define as particular amounts of a substance surrounded by a boundary over which energy may pass only in the form of heat or work. An example of a system is a mass of gas or vapour contained in an engine cylinder, the boundary of which is drawn by the cylinder walls, the cylinder head and the piston crown (the valves are assumed to be closed). Such a system is called a CLOSED SYSTEM, for the boundary is continuous and no matter may enter or leave. Energy may, however, pass across the boundary of a system, and when this happens the energy is

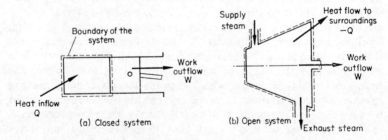

Fig. 1.

always in the form of heat or work. For example, the cylinder head might be heated with the valves closed and heat flow occur through the cylinder head into the gas in the cylinder, and the resulting pressure of the gas might move the piston against some resistance so that work flows outwards from the system. Notice that the system remains closed even if the piston moves, for the substance does not move across the boundary.

A system may also exist within a boundary which has breaks in it through which matter may enter and leave—such an arrangement is called an OPEN SYSTEM. A steam turbine is a good example of an open system, for there is an entrance through which high pressure steam can be supplied and an exhaust port through which low pressure steam may leave after its passage through the turbine. The boundary of the open system is formed by the turbine casing, and again if energy crosses the boundary it must be in the form of either heat or work. Thus heat may flow to the surroundings (although steps are taken to prevent it) and work may flow outwards along the rotor shaft to drive a generator.

ENERGY, HEAT AND WORK

We have seen that energy can exist in various forms. In this subject two main groups of energy are of importance.

(a) Stored Energy. This refers to those forms of energy which can be stored in a system, such as internal energy, potential energy and kinetic energy. Thus a system might consist of a fluid at some height z above a datum line, moving at velocity v and being at some temperature T. The system will possess an amount of stored energy equal to the sum of its potential energy due to its height, its kinetic energy due to its velocity and its internal energy due to its temperature. There are other forms of stored energy such as chemical energy, electrical energy and nuclear energy which will not concern us in this subject.

(b) Transitional Energy. This refers to energy which is being transferred between a system and its surroundings. Both heat and work are forms of energy which is being transferred and are the only forms in which energy can cross the boundaries of a system. Neither heat nor work can exist as stored energy, so that it is incorrect to speak of a body 'containing heat' or 'containing work'. A system may contain internal energy and possess both potential and kinetic energy, but heat or work energy only exist on their way to or from a system.

Consider the analogy of rain falling into a reservoir of water. When the water is falling it is called 'rain' but as soon as it enters the reservoir it becomes additional water and no longer exists as rain. Heat and work are analagous to rain—as soon as they cross the boundary into a system they cease to exist as heat or work but add to the stored energy of the system in the form of potential, kinetic or internal energy. If heat and work enter a system the sum of stored energy is increased, and if heat and work leave a system the sum of stored energy is decreased.

3

From the above, since heat and work are both forms of transitional energy we must expect them to be related to each other. Work is measured as the product of the average force exerted and the distance it moves along its line of action; the units of work are therefore newton-metres (= joules). Heat is also measured in joules.

Dr. James Joule (1818–1889) conducted a series of experiments in which he allowed falling weights to drive rotating paddles in a calorimeter filled with water, thus causing work to flow into the water. The result was to increase the temperature and hence the internal energy of the water. The same result could, of course, have been obtained by heating the water in the calorimeter. Joule found that there was a mechanical equivalent of heat. The S I system of units recognises this equivalence by making the joule (1 newton × 1 metre) the unit of both heat and work.

PROPERTIES OF A SYSTEM

A system possesses a number of properties which are used to describe the *state* of the system. A property is a characteristic of a system which depends upon the state of the system *but not upon how that state was reached*. Thus, for example, the pressure of a gas which forms a system is one of its properties, and the pressure measurement helps to describe the state of the gas; but it does not matter how the particular pressure was arrived at.

The state of a system is fully defined by two of its properties, providing the properties are independent of each other. Care, however, must be taken to ensure that the properties chosen are independent ones. For example, pressure and temperature are two independent properties of a gas and completely define its state. On the other hand, as we shall see later on in our work, these properties of pressure and temperature are not independent of each other for a vapour (a mixture of gas and liquid). In the case of a vapour, the state is fully defined by the independent properties of pressure and specific volume. We shall examine each of the properties of temperature, pressure and specific volume in turn and see what is meant by them, how they are measured, and then later on in the book we shall observe their importance in the working of heat engines.

TEMPERATURE

When heat or work energy flows into a closed system, the internal energy of the system increases and this is usually (but not always) manifested as a rise in its temperature. Measurement of temperature

4

is made by some form of thermometer. The basic S I unit of temperature is the degree Kelvin (K) and the derived unit is the degree Celsius or degree Centrigrade (degC).

(a) **The Celsius or Centigrade Scale.** In this scale 1 degC is $\frac{1}{100}$ of the rise in temperature caused by heating water from its freezing-point (called 0°C) to its boiling point (called 100°C) at a constant pressure of 101·325 kN/m².

In the Celsius scale, the zero point (0°C) was chosen to be that of melting ice. *There is, however, a certain degree of coldness beyond which no lower temperature is possible.* This occurs at about −273°C, and is called the ABSOLUTE ZERO.

If a given volume of air is kept at constant pressure, and its temperature is reduced by 1 degC, experiment shows that its volume reduces by about $\frac{1}{273}$ of its volume at 0°C. Thus, if 273 m³ of gas be reduced from 0°C to −1°C, the volume becomes 272 m³; at −2°C it becomes 271 m³ and so on.

	Temperature		
	0°C − volume		273 m³
	−1°C −	,,	272 m³
	−2°C −	,,	271 m³
and theoretically at	−273°C −	,,	0 m³

In fact, the air, or any other gas, would first liquefy and then become solid before the absolute zero was reached.

(b) **The Kelvin (Absolute) Scale.** This scale of temperature uses the fundamental basis of absolute zero for the beginning of the scale, instead of the arbitrary zero of the centigrade scale. Now the behaviour of gases, which of course we must study since they are the working agent in many engines, depends on their real or absolute temperature and not on any artificial value expressed on some arbitrary scale. All fundamental formulae are expressed in absolute temperatures (T), and the student should be able to convert temperatures from one scale to the other.

Conversion from °C to K (degree Kelvin or °C absolute)

$$T\,\mathrm{K} = 273 + t\,°\mathrm{C}$$

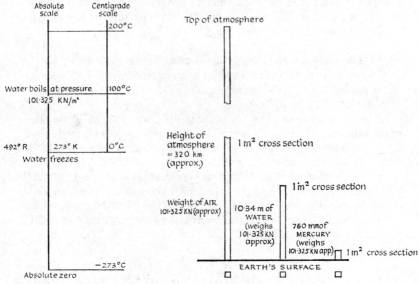

Fig. 2a. Temperature Scales   Fig. 2b.   Pressure Scales

PRESSURE

(a) **Atmospheric Pressure.** This is due to the weight of the column of air above the earth's surface. The pressure on one square metre of the earth's surface is equal to the weight of the column of air of one square metre cross section above where the measurement is being made. It may also be expressed as the height of a column of liquid of equal cross section which weighs the same as the air column. The atmospheric pressure varies slightly from day to day and is recorded by a barometer, but usual values are:—

$$101 \cdot 325 \text{ kN/m}^2 = 10 \cdot 34 \text{ m of water} = 760 \text{ mm of mercury}$$

It is necessary to be able to convert easily from one unit of pressure to another, and the following constants are worth remembering.

1 m³ of water has a mass of 1000 kg and will exert 9·81 kN

$$\therefore 1 \text{ m of water} = 9 \cdot 81 \text{ kN/m}^2$$

Mercury is 13·6 times as heavy as water

$$\therefore 1 \text{ m of mercury} = 13 \cdot 6 \text{ m of water}$$
$$= 13 \cdot 6 \times 9 \cdot 81$$
$$= 133 \cdot 3 \text{ kN/m}^2$$
$$1 \text{ mm of mercury} = \underline{133 \cdot 3 \text{ N/m}^2}.$$

6

Gauges which record pressures in lbf/in² are likely to be around for a considerable time.

To convert from one to the other the relationship

$$1 \text{ lbf/in}^2 = 6897 \text{ N/m}^2 \text{ is useful}$$

e.g.

A pressure gauge reads 110 lbf/in². What is the pressure in S I units?

$$110 \text{ lbf/in}^2 = 110 \times 6897$$
$$= 759 \text{ kN/m}^2$$

(b) Gauge Pressure and Absolute Pressure. Pressure in a pipeline or a vessel is normally measured above the atmospheric pressure by some sort of pressure gauge. For instance, the manometer gauge shown in Fig. 3 is being used to measure the pressure of

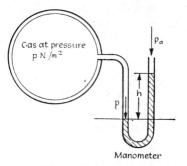

Fig. 3. Manometer Gauge

gas in a pipeline. Suppose the gauge to give a reading of h mm difference between the water levels in the two limbs of the U tube. The pressure of the gas p is being balanced by the atmospheric pressure p_{at} plus the pressure due to the column of water h. Hence

$$p = p_{at} + h.$$

i.e. the true or absolute pressure of gas in the pipeline is the sum of the atmospheric pressure and the gauge reading.

Absolute pressure = gauge pressure + atmospheric pressure.

Example 1

'The pressure of the gas supplied to an engine is measured as 64 mm of water gauge. If the barometer reads 750 mm of mercury, what is the absolute pressure of the supply gas in kN/m²?'

7

Gauge pressure $p_g = 64$ mm of water

$$= \frac{64}{13 \cdot 6} \text{ mm of mercury}$$

Absolute pressure $(p) =$ gauge pressure (p_g) + atmospheric pressure (p_{at})

$$p = \frac{64}{13 \cdot 6} + 750 \text{ mm of mercury}$$

$$= \left(\frac{64}{13 \cdot 6} + 750 \right) 133 \cdot 3$$

$$= 101 \cdot 3 \text{ kN/m}^2.$$

When recording pressures, and indeed, when recording the readings of all other instruments used in routine engineering measurement, it is pointless to produce an answer showing an accuracy greater than that which the instrument itself is capable of reading. For example, if the steam pressure gauge on a boiler reads $0 \cdot 5$ MN/m^2 when the barometer reads 752 mm of mercury, it is pointless to record

Absolute pressure $= p_g + p_{at}$

$$= 0 \cdot 5 + \frac{752 \times 133 \cdot 3}{10^6}$$

$$= 0 \cdot 5 + 0 \cdot 0902$$

$$= 0 \cdot 5902 \text{ MN/m}^2,$$

when the pressure gauge itself is graduated at intervals of 25 kN/m^2! The answer must bear a proper relation to the accuracy of the instruments used, and in this case Absolute pressure $= 0 \cdot 59$ MN/m^2 is quite good enough.

S.T.P. AND N.T.P.

Scientists and engineers do not agree on the definitions of 'standard temperature and pressure' (s.t.p.) and 'normal temperature and pressure' (n.t.p.). In this book the following interpretations are used:—

Standard temperature and pressure is a set of standard conditions used for reference purposes, being a pressure of $101 \cdot 325$ kN/m^2 and a temperature of 0°C.

Normal temperature and pressure refers to the atmospheric conditions of a normal day, namely a pressure of 101·325 kN/m² and a temperature of 15°C.

SPECIFIC VOLUME

The specific volume of a substance is its volume per unit mass and we shall normally measure specific volume in m³/kg. As the temperature of a substance rises it increases in volume if free to do so. (There is a peculiar exception from this general rule in the case of water which has a minimum specific volume at about 4°C but we shall not be concerned with that here.) The proportionate expansion is much greater when the system is in the gaseous state than when it is in either the liquid or solid state. The *density* of a substance, which is defined as the mass per unit volume, is the inverse of specific volume

$$\text{i.e. density } \rho = \frac{1}{V}$$

FLOW AND NON-FLOW PROCESSES

A process occurs when a system undergoes a change of state, and such a process can be shown on a state chart by a line which represents the progressive change of state. Thus a line on a pressure-volume (pV) diagram may represent the progressive change of state of a gas system, and a point on the line represents the state at any particular part of the process. The line on the pV diagram in Fig. 4 shows the way the state of a gas might change on being heated from state 1 to state 2 so that its pressure increases from p_1 to p_2 and its volume increases from v_1 to v_2.

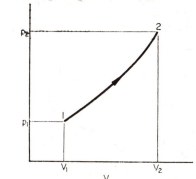

Fig. 4. pV Diagram

9

A process may be NON-FLOW in which a fixed mass within a defined boundary is undergoing a change of state: for example, a gas which is being heated in a closed cylinder undergoes a non-flow process. The process is non-flow although the cylinder might be fitted with a movable piston, for no substance crosses the boundary. Closed systems undergo non-flow processes.

A process may be FLOW in which mass is entering and leaving through breaks in the boundary of an open system. In a *steady-flow process* mass flows steadily across the break in the boundary from the surroundings at entry, and an equal mass flow steadily across the break in the boundary at exit so that the total mass of the system at any time remains constant. For example, in a gas turbine engine, air flows into a compressor which raises its pressure, after which fuel is added and the mixture ignited in a combustion chamber. The resulting hot gases are passed through a turbine where work is done on the blades and the gases are finally exhausted into the atmosphere. There is a flow of substance (in this case air plus a little fuel) through the machine and the system is said to have been conducted through a flow process.

It is often useful to consider part only of a flow process, in which case the part considered may be a non-flow process. For example, if the compression process of the mixture in the cylinder of a reciprocating petrol engine is considered separately, the process is non-flow, although the whole action of the engine which takes in air and fuel and rejects exhaust gas is a flow process.

CYCLE OF OPERATIONS

A closed system may pass through a series of processes and return to its initial state, in which case it is said to have passed through a *cycle*. The processes through which the system has passed may be shown on a state diagram.

In Fig. 5 a system originally at state 'a', measured by its proper-

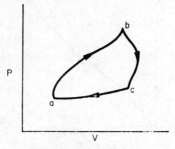

Fig. 5. State Diagram

ties of pressure p_a and volume V_a, has passed through a cycle via the processes a–b, b–c and c–a, and has returned to its original state 'a' where its properties are the same as it started with.

Suppose, for example, a system consisted of some gas in a cylinder fitted with a movable piston. If the gas was heated its pressure would rise so that the gas could force the piston outwards as it expanded. The gas might then be cooled and the piston driven in again to the point at which the gas is returned to its original volume and pressure. The system would have been conducted through a cycle, similar in some ways to the cycle carried out in an engine.

SPECIFIC HEAT CAPACITY

Experiments with different substances show that they require different amounts of heat to effect the same increase in temperature, and specific heat capacity may be defined as the amount of heat required to raise unit mass of the substance through one degree. The units are therefore joule/kgK.

When a body is heated, the heat energy is used to increase the internal energy by speeding up the molecules and also to provide the work necessary to expand the body. In a solid or a liquid, the amount of expansion is very small and the work of expansion is similarly small. When heating a gas, however, the expansion may be considerable, and a gas has two values of specific heat capacity as follows:—

(a) Specific Heat Capacity at Constant Volume (c_v). Consider 1 kg of gas being heated in a closed container, so that no expansion of the gas is allowed. The number of joules required to raise the temperature of 1 kg of the gas through 1 degK under these conditions is called the Specific Heat Capacity at Constant Volume.

In this case there is no work of expansion, because the gas cannot move (Work = Average Force × Distance moved), and all the heat supplied is used to increase the internal energy of the gas. The value for c_v for air is approximately 710 J/kgK.

Fig. 6. Specific Heat at Constant Volume

11

(b) Specific Heat Capacity at Constant Pressure (c_p). Consider now 1 kg of gas being heated in a cylinder fitted with a movable piston, which exerts a constant pressure on the gas. When the gas is heated it will expand and move the piston through some distance h. In this case, in addition to the heat required to increase the internal energy of the gas, further heat must be added to perform the work of moving the piston through the height h.

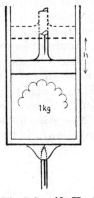

Fig. 7. Specific Heat at Constant Pressure

The number of joules required to raise the temperature of 1 kg of the gas through 1 degK under these conditions is called the Specific Heat Capacity at Constant Pressure.

The value of the specific heat of a gas at *constant pressure will always be greater than that at constant volume* by the amount of expansive work done. The value of c_p for air is approximately 1005 J/kgK.

The ratio $\dfrac{c_p}{c_v}$ is frequently used in gas calculations and for convenience is written as γ

$$\frac{c_p}{c_v} = \gamma$$

THE ENGINE MECHANISM

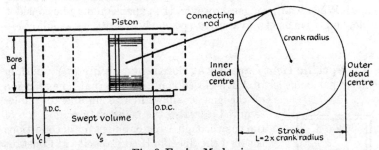

Fig. 8 Engine Mechanism

In the engine mechanism, the reciprocating motion of the piston is converted into the rotary motion of the crank. The length of travel of the piston between its extreme positions is called the 'Stroke', and is equal to twice the crank radius.

The total volume swept through by the piston during a single stroke is called the SWEPT VOLUME (V_s).

12

$$V_s = \text{Cross Sectional Area of piston} \times \text{stroke}$$

$$= \frac{\pi d^2}{4} \times L \quad \text{where } d = \text{bore}$$
$$\text{and } L = \text{stroke}$$

When the piston is at the top of its stroke, there remains a volume enclosed between the piston crown and the cylinder head, called the CLEARANCE VOLUME (V_c).

The ratio between the maximum and minimum volumes of a cylinder is called the COMPRESSION RATIO (r), i.e. the ratio in which a gas would be compressed if occupying the total cylinder volume initially and the clearance volume finally, which would occur as the piston moved between its extreme positions.

$$\text{Now, Maximum cylinder volume} = V_c + V_s$$
$$\text{Minimum cylinder volume} = V_c$$

$$\therefore \text{Compression ratio } r = \frac{V_c + V_s}{V_c}$$

The CAPACITY of any engine is the sum of the swept volumes of all its cylinders.

CALORIFIC VALUE OF A FUEL

A fuel may be defined as a substance which, when ignited in the presence of a suitable supply of oxygen, will give out an appreciable quantity of heat. The calorific value (c.v.) of a fuel is the name given to its heating value, and states the amount of heat released during complete burning of unit quantity of the fuel. The calorific value of solid or liquid fuels is usually expressed in joules per kg, and of gaseous fuels in joules/m³ at n.t.p.

FUEL CONSUMPTION OF AN ENGINE

The amount of fuel used by an engine may be expressed in kg per hour for solid or liquid fuels, or m³ per hour for gaseous fuels. A more useful definition is that of SPECIFIC FUEL CONSUMPTION, which states the amount of fuel used by the engine when running for 1 hour, whilst developing a constant output of 1 kilowatt. Thus, specific fuel consumption states not only how much fuel the engine is using, but also what output one is getting for it. The units of s.f.c. are kg/kWh (or m³/kWh).

13

BRAKE POWER (WATTS), INDICATED POWER (WATTS), AND MECHANICAL EFFICIENCY

Power is defined as the rate of doing work, and a work-rate of one joule per second is named the WATT.

The brake power (b.p.) of an engine is the power available for work at the output shaft of the engine, and is measured using some form of brake.

The indicated power (i.p.) of an engine is the power developed in its cylinders, and is measured by a form of pressure indicator connected to the cylinder head. The indicated power is always greater than the brake power of an engine, because there will always be a reduction of power between the cylinder and the output shaft due to friction between the moving parts and the pumping power required to charge the cylinder.

$$\text{i.p.} = \text{b.p.} + \text{friction power}$$

The Mechanical Efficiency of an Engine (η_m) is the ratio $\dfrac{\text{b.p.}}{\text{i.p.}}$.

It measures the efficiency with which an engine converts the power developed in its cylinders into useful power available at the output shaft. An average value of mechanical efficiency for a petrol engine running at normal speed would be 0·80, or 80%.

THERMAL EFFICIENCY (η_t)

An engine cannot convert all the heat energy of the fuel into work, for reasons which will be described later (Chapter 5). The thermal efficiency of an engine is the measure of its ability to do so, and is written

$$\eta_t = \frac{\text{Work done per second}}{\text{Heat supplied from the fuel per second}}$$

Thermal efficiency can be based on either i.p. or b.p. The former is preferable, since it will not include the inefficiency due to friction losses, but as many engines are not normally fitted with indicators for measuring i.p., the thermal efficiency based on b.p. is also acceptable.

Indicated Thermal Efficiency

$$\text{indicated } \eta_t = \frac{\text{Work done per second (watts} = \text{joules/second)}}{\text{Heat supplied per second}}$$

Heat supplied per second = kg of fuel supplied/second × calorific value of fuel (joules/kg).

$$\therefore \eta_t = \frac{\text{i.p. (watts)}}{\text{kg of fuel/second} \times \text{c.v.}}$$

> Alternatively:—
>
> $$\text{indicated } \eta_t = \frac{\text{Work done/hour}}{\text{Heat supplied/hour}}$$
>
> $$= \frac{\text{i.p.} \times 3600}{\text{kg of fuel/h} \times \text{c.v.}}$$

but $\dfrac{\text{kg of fuel/hour}}{\text{indicated power in kW}} = $ Specific fuel consumption (kg/kWh)

$$\therefore \eta_t = \frac{3600 \times 10^3}{\text{Specific fuel consumption} \times \text{c.v.}}$$
$$(\text{kg/kWh})$$

Similarly:

Brake Thermal Efficiency

$$\text{brake } \eta_t = \frac{\text{b.p. (watts)}}{\text{kg of fuel/second} \times \text{c.v.}}$$

Example 2

'An oil engine developing 37·5 kW uses 9·0 kg of oil per hour of calorific value 45 MJ/kg. 8·5 kW are used to overcome friction in the engine itself. Determine

(a) b.p.
(b) Specific Fuel Consumption kg/kWh (brake basis).
(c) Mechanical Efficiency.
(d) Indicated Thermal Efficiency of the engine.'

(a) b.p. = i.p. – friction power
$$= 37 \cdot 5 - 8 \cdot 5 = \underline{29 \cdot 0 \text{ kW}}$$

(b) Fuel Consumption = 9·0 kg/h

$$\therefore \text{ Specific Fuel Consumption} = \frac{9}{29} \text{ kg/kWh}$$

$$= \underline{0 \cdot 310 \text{ kg/kWh}}$$

15

(c) Mechanical Efficiency $= \dfrac{\text{b.p.}}{\text{i.p.}}$

$$= \frac{29 \cdot 0}{37 \cdot 5} = 0 \cdot 773 \text{ or } 77 \cdot 3\%$$

(d) Indicated Thermal Efficiency $= \dfrac{\text{i.p. (watts)}}{\text{kg of fuel/sec} \times \text{c.v.}}$

$$= \frac{37 \cdot 5 \times 10^3 \times 3600}{9 \times 45 \times 10^6}$$

$$= 0 \cdot 333 \text{ or } 33 \cdot 3\%$$

MOLECULAR WEIGHT

A molecule may be defined as the smallest quantity of a substance which can exist independently and retain the characteristics of that substance. A molecule may consist of one or more atoms, the number of atoms per molecule of a substance being shown in the chemical formula by an index at the base of the symbol. Thus, for example,

C (Carbon) has one atom per molecule. All such substances are called 'monatomic'.

H_2 (Hydrogen) has two atoms per molecule. All such substances are called 'diatomic'.

CO_2 (Carbon Dioxide) has three atoms per molecule, one of carbon and two of oxygen. All such substances are called 'triatomic'.

CH_4 (Methane) has five atoms per molecule, one of carbon and four of hydrogen. All substances with more than three atoms per molecule are called 'polyatomic'.

The actual weight of any atom is very small, and it is usual to express atomic weights in terms of the lightest known atom, that of hydrogen, e.g. the atomic weight of hydrogen is 1.*

The atomic weight of oxygen is 16, which states that oxygen is 16 times as heavy as hydrogen.

A table is given of some common atomic and molecular weights which will be useful in dealing with problems on combustion:—

---

* Recent work has shown that the atomic weight of hydrogen is slightly greater than unity, and it is now customary to express atomic weights in relation to that of oxygen which is taken as 16.

16

Substance	Chemical Symbol	Atomic Weight	Number of Atoms per Molecule	Molecular Weight
Hydrogen	H₂	1	2	2
Carbon	C	12	1	12
Nitrogen	N₂	14	2	28
Oxygen	O₂	16	2	32
Sulphur	S	32	1	32

CONSERVATION OF ENERGY—THE FIRST LAW OF THERMODYNAMICS

Our experience shows that energy cannot be created or destroyed, although it can be changed from one form to another, and this is known as the principle of *Conservation of Energy* * For example, although it would appear that when a moving vehicle is stopped its kinetic energy is destroyed, in fact the energy flows first of all as heat to the brakes and later on from the brakes to the surroundings as they cool down. Exactly the same amount of heat energy is created by friction at the brakes as was possessed by the vehicle as kinetic energy when it was moving.

Some kinds of energy are more easily changed than others. Thus, it is easy to convert kinetic energy into internal energy through the medium of friction (i.e. work), but it is much more difficult to convert the heat energy released from a fuel into work. Stone Age man was able to produce fire by rubbing two sticks together, but it was not until about the year 1700, when the first steam engine was produced, that the reverse procedure of releasing heat energy from a fuel and changing it into work was achieved, and even now we can only convert about one-third of a given flow of heat energy into work.

The interchange of energy between its various forms is a matter of primary concern to an engineer. The *first law of thermodynamics* refers to such changes of energy and expresses the principle of conservation of energy thus, 'Given a total amount of energy in a system and its surroundings, this total remains the same whatever changes of form might occur'. From this law, it follows that when an engineer tries to convert one type of energy to another, he can measure the success of his efforts by drawing up an account of how the energy he had to start with has been shared out. For example, in a heat engine test the following rather simplified account might result:—

* The principle of Conservation of Energy as expressed here, whilst accurate enough for the purpose of our work in heat engines, is far from being a complete statement. We know, for example, that energy can be created from mass in an atomic reactor, and more recently the theory of the continuous creation of matter and energy in space has been put forward.

Heat energy released from the fuel flowing into the system (100)

↗ Work flowing from the system.   30

→ Heat flow to the engine cooling water   30

↘ Heat flow to the exhaust gases and surroundings   40

We shall now discuss the application of the first law of thermodynamics to processes in closed and open systems.

(a) For a closed system being conducted through a non-flow process. Consider a closed system which is being conducted through a non-flow process during which there is a heat flow Q into the system and a work flow W from the system. The first law states, 'The difference between the sum of the heat flowing into a closed system and the work flowing from the system is equal to the increase in the internal energy of the system'.

Thus: If work flows from a system which receives no heat, it does so at the expense of the internal energy of the system.

If a system receives an inflow of heat, and no work flows from the system, then its store of internal energy is increased.

If a system receives an inflow of heat, and at the same time some work flows from the system, then its internal energy will increase or decrease by the amount by which the quantity of heat flowing in is greater or less than the quantity of work flowing out. This is expressed mathematically in the energy equation:—

$$Q \qquad\qquad W \qquad\qquad (U_2 - U_1)$$

heat flow into the system	$=$	work flow from the system	$+$	increase of internal energy of the system

If Q is positive, heat is flowing into the system,

Q is negative, heat is flowing outwards from the system,

W is positive, work is flowing from the system,

W is negative, work is flowing into the system,

$(U_2 - U_1)$ is positive, the internal energy of the system is increasing,

$(U_2 - U_1)$ is negative, the internal energy of the system is decreasing.

Suppose, for example, that the gas in the clearance volume of an engine is expanding by forcing the piston outwards against an external resistance through the length of the stroke. Assume that during the same time 1·0 kJ of heat flows into the gas from the cylinder head. Given that the force resisting the motion of the piston is 700 N and the length of stroke is 0·5 m, what is the change in internal energy of the gas during the process?

$$\frac{\text{Work done by the}}{\text{expanding gas}} = \text{resisting force} \times \text{distance moved by the piston}$$

$$= 700 \times 0.5$$
$$= 350 \text{ Nm} = 350 \text{ J}$$

(and this is the work which flows from the system).
Since the system is closed:—

$$Q = W + (U_2 - U_1),$$

$\therefore$ Increase of Internal Energy $(U_2 - U_1) = Q - W$
$$= 1000 - 350$$
$$= 650 \text{ J}$$

(b) For a closed system which has passed through a cycle.
When a closed system has passed through a cycle then its state at the end of the cycle is the same as it was when the cycle started. During the series of non-flow processes occurring during the cycle the system will have received and delivered heat and work across its boundaries. Now since the end states are identical this means that the properties of the system are unchanged, so that the internal energy of the system (which is one of its properties) is unchanged. Thus none of the heat or work has been used to increase or decrease the internal energy of the system, and the first law states in this case, *when a closed system has passed through a cycle, the sum of the heat energy taken in across the boundary from the surroundings is equal to the work delivered from the system to the surroundings.*
i.e. for a closed cycle

$$\frac{Q}{\substack{\text{Heat flow into the} \\ \text{system}}} = \frac{W}{\substack{\text{work flow from the} \\ \text{system}}}$$

(c) For an open system undergoing a steady flow process.
Fig. 9 represents a system in which a steady-flow process is taking place.

The working fluid is assumed to be passing through the region defined by the boundaries at a steady rate so that mass is leaving the system at the same steady rate as it enters. It is important to recognise that, in addition to the work W shown flowing from the system, work will need to be done by the surroundings on the system to cause the fluid to enter, and similarly the fluid leaving the system will do work *on* the surroundings. This flow-work, as it is called, exists only when the fluid is flowing, and is required to push the existing fluid out of the way to make room for the incoming or outgoing fluid.

19

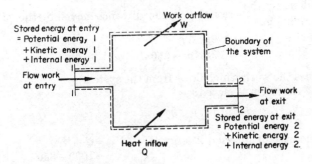

Fig. 9. Flow Process in an Open System

The first law states that

stored energy at entry + flow-work at entry + heat inflow to the system
= stored energy at exit + flow-work at exit + work outflow from the
system.

TRANSMISSION OF HEAT

Heat may flow in three ways:—

By Conduction. We have already observed that the hotter a body is, the greater is the kinetic energy of its molecules. Consider a metal rod having one end at a higher temperature than the other. The molecules at the hotter end will be moving faster than those at the colder end. By impacts between adjacent molecules their motion will be transferred along the rod, and energy is thereby transferred. This process is known as Heat Transmission by CONDUCTION.

By Convection. When heat is applied to a fluid, that part of the fluid nearest to the source of heat will expand and so become less dense. This lighter section of the fluid will rise and its place will be taken by heavy fluid. The process is repeated and 'convection' currents are set up, heat being transmitted by physical movement of the fluid. This process is known as Heat Transmission by CONVEC-TION.

By Radiation. Heat may be transmitted from a hotter to a colder body by wave motion, the waves being of the same nature as those of light or radio. This process is known as Heat Transmission by RADIATION and, in contrast to conduction or convection heat transmission which require a medium, radiation can take place in a

20

vacuum (e.g. radiated heat is received from the sun through the space vacuum).

Most heat flow takes place as a combination of two or three of the above methods, e.g. the air in a room close to a fire receives heat from the fire by both radiation and conduction and, being less dense than the air further from the fire, rises and transmits heat throughout the room by convection.

REVISION EXERCISES—CHAPTER 1

1. A petrol engine has a cylinder diameter of 62 mm, stroke 95 mm and compression ratio 6·4. Calculate the clearance volume.

(Ans. 53·1 cm³)

2. An oil engine developes 96 kW at 360 rev/min, and uses 27 kg of oil per hour of c.v. 42·84 MJ/kg. If the friction power is 24, find the i.p., the Mechanical Efficiency and the Brake Thermal Efficiency.

(Ans. 120, 80·0%, 29·8%)

3. In an instrument used for the measurement of air flow in a pipeline, a manometer shows the pressure in the pipe to be 8 cm of water below the atmospheric pressure. What is the absolute pressure of the air in kN/m² in the pipeline if the barometer reads 750 mm of mercury?

(Ans. 99·22 kN/m²)

4. A car driven by a petrol engine travels at 50 km/h when developing 17·15 kW. If the thermal efficiency is 25% and the c.v. of the petrol is 43 MJ/kg, determine the specific fuel consumption of the engine in kg/kWh, and the cost of fuel per kilometre. Take the cost of petrol as 6p per litre, and specific gravity of petrol 0·72.

(Ans. 0·335, 0·96p)

5. The engine of a jet-propelled aircraft develops 750 kW with an overall thermal efficiency of 10% when using kerosene oil as fuel. What mass of oil of c.v. 43 MJ/kg will be required for a flight of 400 km at an average speed of 800 km/h?

(Ans. 313·2 kg)

6. A gas expands in a closed cylinder and exerts a constant pressure of 420 kN/m² on a 50 mm diameter piston as it moves through a distance of 150 mm. Determine the heat flow which must take place into the system during the process so that its internal energy shall remain the same.

(Ans. 123·7 J)

2. The Gas Laws

Many heat engines obtain their power by using the pressure of expanding gases to drive forward a piston in a cylinder. It is of first importance, therefore, to make a study of how gases behave, since the amount of power available at the output shaft of any engine depends on obtaining the best possible performance from the gases in the cylinder.

We must first refer to two experimental laws which state the relationship between the conditions of temperature, volume and pressure of a gas. These laws are usually stated with reference to some imaginary 'perfect gas', imaginary because it is supposed to follow the laws exactly, which no known gas does. Nevertheless, the laws can be applied to real gases such as exist in an engine cylinder, and will give sufficiently accurate results for all practical purposes.

BOYLE'S LAW

Boyle found by experiment that when a given mass of gas is kept at constant temperature, its volume varies inversely as its *absolute* pressure, e.g. double the absolute pressure gives half the volume, three times the absolute pressure gives one-third of the volume, and so on.

$$\text{i.e. } V \propto \frac{1}{p}$$

$$\text{or } pV = \text{constant}$$

CHARLES' LAW

Charles found experimentally that when a given mass of gas is kept at constant pressure, its volume varies directly as its *absolute* temperature, e.g. at double the absolute temperature, the volume is doubled. At three times the absolute temperature, the volume is trebled, and so on.

$$\text{i.e. } V \propto T$$

$$\text{or } \frac{V}{T} = \text{constant}$$

THE CHARACTERISTIC GAS EQUATION

Consider a volume V_1 of a gas at pressure p_1 and temperature T_1. Let the gas expand to some intermediate volume V whilst the temperature remains constant at T_1.

From Boyle's Law: $p_2 V = p_1 V_1$ and hence the resulting pressure

$$p_2 = \frac{p_1 V_1}{V} \qquad \qquad \cdots [1]$$

Now let the gas be heated to some temperature T_2 whilst the pressure remains constant at p_2 and the volume increases to V_2.

From Charles' Law $\dfrac{V}{T_1} = \dfrac{V_2}{T_2}$ and hence the resulting volume

$$V_2 = V \frac{T_2}{T_1}$$

$$\text{i.e. } V = \frac{V_2 T_1}{T_2} \qquad \qquad \cdots [2]$$

Substituting for V in equation [1]

$$p_2 \cdot \frac{V_2 T_1}{T_2} = p_1 V_1$$

$$\text{i.e. } \frac{p_2 V_2}{T_2} = \frac{p_1 V_1}{T_1}$$

Hence we may write, for a perfect gas,

$$\frac{pV}{T} = \text{constant}$$

or, since a constant mass of gas was considered,

$$\frac{pV}{T} = m \times \text{constant}$$

$$\text{or } \frac{pV}{T} = mR$$

which is the Characteristic Gas Equation.

23

THE CHARACTERISTIC GAS CONSTANT

R is called the 'Gas Constant' and is usually expressed in units of J/kgK. Taking these units for R, it is essential to express the other items in the equation in the corresponding units, which are

p = absolute pressure of the gas [N/m²]
V = volume of the gas [m³]
T = absolute temperature of the gas [$(t + 273)$ K]
m = mass of gas [kg]
R = constant [J/kgK]

Notice that since $\dfrac{pV}{T}$ = constant, we may write

$$\frac{p_1 V_1}{T_1} = \frac{p_2 V_2}{T_2} = \underline{\qquad} \text{ to connect the conditions of a gas}$$

at any stage in a non-flow process.

Example 3

'The pressure of the gas supplied to an engine is measured as 75 mm water gauge when the barometer reads 760 mm of mercury. Determine the volume of 1 kg of the gas at this pressure and a temperature of 85°C if R for the gas is 287 J/kgK.'

$$\text{Pressure } p = \left(760 + \frac{75}{13 \cdot 6}\right) \text{ mm of mercury}$$

$$= \left(760 + \frac{75}{13 \cdot 6}\right) 133 \cdot 3 \text{ N/m}^2$$

$$= 102 \cdot 1 \text{ kN/m}^2$$
$$= (85 + 273) \text{ K} = 358 \text{ K}$$
$$\frac{pV}{T} = mR$$

$$\therefore \ V = \frac{mRT}{p} = \frac{1 \times 287 \times 358}{102 \cdot 1 \times 10^3}$$

$$= 1 \cdot 007 \text{ m}^3$$

Example 4

'A gas at a temperature of 17°C and pressure of 125 kN/m² occupies a volume of 2·46 m³. If the gas be now compressed to a volume of $1 \cdot 0$ m³ and pressure of 700 kN/m², what would be its temperature?'

$$\frac{p_1 V_1}{T_1} = \frac{p_2 V_2}{T_2} \qquad T_1 = (17 + 273) \text{ K}$$

$$= 290 \text{ K}$$

i.e. $\dfrac{125 \times 10^3 \times 2 \cdot 46}{290} = \dfrac{700 \times 10^3 \times 1}{T_2}$

$$\therefore T_2 = \frac{700 \times 290}{125 \times 2 \cdot 46}$$

$$= 660 \cdot 3 \text{ K}$$

$\therefore$ final temperature $t_2 = (660 \cdot 3 - 273)°C$

$$= 387 \cdot 3°C$$

Example 5

'A balloon of capacity 500 m³ is inflated at ground level where the pressure is 100 kN/m² and temperature 15°C, with 125 m³ of hydrogen at a temperature of 15°C and pressure 300 kN/m². Taking R for air $= 287$ J/kgK, and R for hydrogen $= 4150$ J/kgK:—

(*a*) Calculate the weight of hydrogen used and the lifting effort of the balloon at ground level.

(*b*) Calculate the pull on the anchor rope when the balloon is at an altitude where the pressure of the atmosphere has fallen to 75 kN/m², and the temperature to $-15°C$.

(*c*) Find the diameter of steel rope required so that the tensile stress in it shall not exceed 75 MN/m². Neglect the weight of the rope.'

(*a*) Mass of hydrogen used $= \dfrac{pV}{RT}$

$$= \frac{300 \times 10^3 \times 125}{4150 \times 288}$$

$$= 31 \cdot 37 \text{ kg}$$

Mass of air displaced at ground level

$$= \frac{100 \times 10^3 \times 500}{287 \times 288}$$

$$= 604 \cdot 9 \text{ kg}$$

$\therefore$ lifting effort $= 9 \cdot 81(604 \cdot 9 - 31 \cdot 37)$ N $= \underline{5 \cdot 627 \text{ kN}}$

(b) Mass of air displaced at upper level

$$= \frac{75 \times 10^3 \times 500}{287 \times 258}$$

$$= 506 \cdot 5 \text{ kg}$$

$$\therefore \text{ Pull on rope} = (506 \cdot 5 - 31 \cdot 37)9 \cdot 81 \text{ N}$$

$$= 4 \cdot 661 \text{ kN}$$

(c) Greatest pull is at ground level. $\text{Stress} = \dfrac{\text{force}}{\text{area}}$

Cross sectional area of rope $= \dfrac{\text{pull at ground level}}{\text{allowable stress}}$

$$\therefore \frac{\pi d^2}{4} = \frac{5627}{75 \times 10^6} \qquad \therefore d = \sqrt{\frac{4 \times 5627}{\pi \times 75 \times 10^6}}$$

$$d = \underline{0 \cdot 009 \ 77 \text{ m}} = 9 \cdot 77 \text{ mm,}$$

$$\text{say 10 mm dia.}$$

AVOGADRO'S HYPOTHESIS

A hypothesis is an assumption of a fact which cannot be proved, but which experience leads one to believe to be true. Avogadro's hypothesis is 'Equal volumes of all gases under the same conditions of temperature and pressure contain equal numbers of molecules'.

Now we cannot prove this statement, for it is not possible to count the vast number of molecules in even a small measurable volume. However, there is plenty of evidence in other directions leading us to believe that the hypothesis is true, and we shall proceed on this assumption.

Consider a number of vessels of equal volume containing samples of different gases at the same temperature and pressure.

All of volume V at pressure P and temperature T

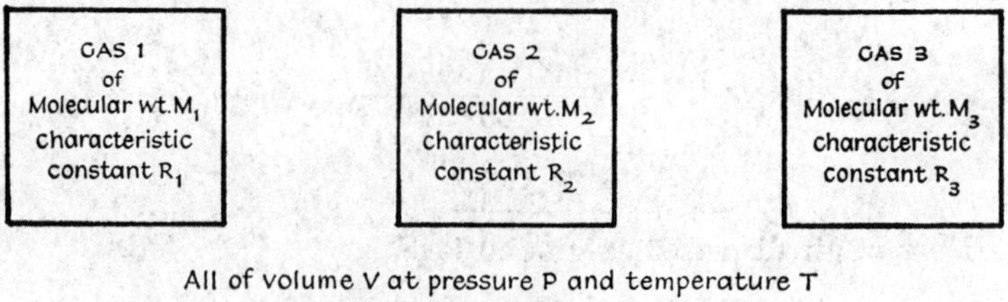

Fig. 10. Avogadro's Hypothesis

According to Avogadro's hypothesis, each vessel will contain an equal number of molecules (say n) of each gas.

$\therefore$ Mass of gas in vessel = number of molecules × mass of each molecule.

26

Applying the Characteristic Equation for the gas in vessel 1

$$\frac{pV}{T} = mR$$

$$\therefore \frac{pV}{T} = nM_1R_1$$

Similarly, for the gases in the other vessels

$$\frac{pV}{T} = nM_2R_2 \text{ and } \frac{pV}{T} = nM_3R_3$$

But, for any gas, $\frac{pV}{T} = $ Constant

$$\therefore \quad M_1R_1 = M_2R_2 = M_3R_3 = \text{Constant}.$$

Now the actual molecular weight of any substance is very small indeed, and the engineer uses the concept of the kilogramme-molecule or kmol which is a mass of the substance numerically equal to the molecular weight in kilogrammes. Thus a kmol of oxygen has a mass of 32 kg, a kmol of hydrogen a mass of 2 kg and so on. Using this conception the constant MR is called 'The Universal Gas Constant' and has the value 8314 J/kmol. Thus, the characteristic constant R for any gas can be determined, if its molecular weight is known, from the expression

$$R = \frac{8314}{M}$$

e.g. R for oxygen $(M = 32) = \dfrac{8314}{32} = 260 \text{ J/kgK}$

R for carbon dioxide $(M = 44) = \dfrac{8314}{44} = 189 \text{ J/kgK}$

Example 6
'The molecular weight of the gas in an engine cylinder is 28·0. If the cylinder bore is 250 mm, the stroke 300 mm, and the compression ratio 5·7 to 1, find the mass of gas in the cylinder if its pressure and temperature at outer dead centre are 200 kN/m² and 160°C respectively. If the gas be now compressed into the clearance volume, what would be the final pressure if the temperature remains the same?'

Swept Volume $V_s = \dfrac{\pi d^2}{4} \times L = \dfrac{\pi}{4} \times 0 \cdot 25^2 \times 0 \cdot 3$

$$= 0 \cdot 014 \ 72 \ \text{m}^3$$

Compression Ratio $\dfrac{V_c + V_s}{V_c} = 5\cdot7$

$$\therefore\ V_c + V_s = 5\cdot7\ V_c$$

$$\therefore\ V_c = \frac{V_s}{4\cdot7} = \frac{0\cdot014\,72}{4\cdot7} = 0\cdot003\,13\ m^3$$

$\therefore$ Total Volume at outer dead centre $= 0\cdot014\,72 + 0\cdot003\,13$

$$V_1 = 0\cdot017\,85\ m^3$$

$$R\ \text{for the gas} = \frac{8314}{28}$$

$$= 297\ J/kgK$$

Mass of gas in cylinder $m = \dfrac{pV}{RT}$

$$\therefore\ m = \frac{200 \times 10^3 \times 0\cdot017\,85}{297 \times 433}$$

$$= 0\cdot0278\ kg$$

For the compression $\dfrac{p_2 V_2}{T_2} = \dfrac{p_1 V_1}{T_1}$, and $T_2 = T_1$ if the temperature remains the same

$$\therefore\ p_2 = \frac{p_1 V_1}{V_2} = 200 \times 10^3 \times 5\cdot7\ N/m^2$$

$$= 1\cdot14\ MN/m^2$$

THE KILOGRAMME MOLE AS A UNIT OF VOLUME

We have described the kmol as a quantity of substance having a mass numerically equal to the molecular weight in kg. Consider 1 kmol of ANY gas at pressure p N/m^2 and temperature T K.

From $\dfrac{pV}{T} = mR$

To find the volume: $V = \dfrac{mRT}{p}$, and since $m = M$ in this case

$$V = \frac{MRT}{p}, \text{ and } MR = 8314\ J/kgK$$

$$\therefore\ V = \frac{8314\,T}{p}$$

28

Hence, if M kg of a gas (i.e. a kmol of the gas) be subjected to a pressure p and temperature T, it will occupy the SAME VOLUME whatever kind of gas is chosen. We may therefore use the kmol as a unit of volume.

For example, at s.t.p. (101·325 kN/m² and 0°C)

$$\text{volume of 1 kmol} = \frac{8314T}{p}$$

$$= \frac{8314 \times 273}{101\cdot325 \times 10^3} = 22\cdot4 \text{ m}^3$$

1 kmol of oxygen (molecular wt 32) has a mass of 32 kg and occupies 22·4 m³ at s.t.p.

1 kmol of hydrogen (molecular wt 2) has a mass of 2 kg and occupies 22·4 m³ at s.t.p.

1 kmol of nitrogen (molecular wt 28) has a mass of 28 kg and occupies 22·4 m³ at s.t.p.

INTERNAL ENERGY OF A GAS

We have seen that the internal energy of a system is the main item of energy stored in it (the other items being potential energy due to its position and kinetic energy due to its movement). It has been explained that internal energy is due to the movement of the molecules within the substance of the system, and that internal energy is a function of temperature. There is no way of measuring the total amount of internal energy a given mass of gas contains, for its behaviour at extremely low temperatures is unknown, but in this study of the behaviour of gases we are only concerned with the *change* in internal energy.

We have already seen (page 19) that when a gas expands in a closed system without taking in or giving out heat and without any work-flow occurring, and hence without change of internal energy, the temperature remains constant. Joule conducted experiments to check this in which a vessel A containing compressed air at approximately 2·3 MN/m² was connected through a valve to a similar vessel B which had been evacuated, the whole being immersed in an insulated bath of water (Fig. 11). When the apparatus had settled to the temperature of the bath, the valve was opened so that the compressed air expanded to fill both vessels. Careful measurement showed no change in the temperature of the water. Since the system was insulated there had been no heat flow, and no work-flow had occurred. From the energy equation for a closed system undergoing a non-flow process

$$Q = W + (U_2 - U_1)$$
$$\text{Nil} = \text{Nil} + (U_2 - U_1)$$

i.e. *there had been no change in Internal Energy*.

On the other hand the pressure of the gas had changed, and the volume had changed, but there was *no change in temperature*. The only condition of the gas which had not changed was temperature, and since the internal energy had not changed Joule inferred that the *internal energy of a given mass of gas depends only on its temperature*.

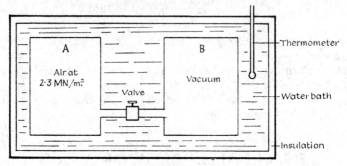

Fig. 11. Joule's Law

Consider m kg of gas being heated in a closed system at constant volume from temperature T_1 to temperature T_2. Since the amount of heat required to raise 1 kg of gas through a degK at constant volume is the specific heat capacity at constant volume c_v

$$Q = mc_v(T_2 - T_1)$$

Since the process occurs at constant volume, no work-flow takes place and applying the energy equation:—

$$Q = W + (U_2 - U_1)$$
thus $$mc_v(T_2 - T_1) = (U_2 - U_1)$$

But, by Joule's reasoning, the internal energy of a given mass of gas depends only on its temperature, so that the change of internal energy depends only on the change of temperature and not on how the change of temperature was brought about. Thus internal energy is a property of a gas, and it follows that for m kg of gas heated from T_1 to T_2 in *any manner*, the change of internal energy is always the same, i.e.

$$U_2 - U_1 = mc_v(T_2 - T_1).$$

Example 7

'10 kg of air is compressed in a closed system from an initial

30

temperature of 30°C to a final temperature of 260°C. Determine the change of internal energy, and if the heat flow from the air to the surroundings during the process is 160 kJ, calculate the work done per kg of air. Take c_v for air = 710 J/kgK.'

Change of Internal Energy $(U_2 - U_1) = mc_v(T_2 - T_1)$
$$\therefore (U_2 - U_1) = 10 \times 710(260 - 30) \text{ J}$$
$$= 1\cdot633 \text{ MJ}$$

Notice that it is not necessary to convert temperatures to absolute values for use in this equation, since
$$(T_2 - T_1) = (t_2 + 273) - (t_1 + 273) = (t_2 - t_1)$$
Heat Flow $Q = -160$ kJ (Q is *negative* since heat is *rejected*)
Now $Q = W + (U_2 - U_1)$
$$\therefore -160 = W + 1633 \text{ kJ}$$
$\therefore W = -1793$ kJ (W is negative, therefore work is done *on* the air)
$$\therefore \text{ Work done on the air per kg} = \frac{1793}{10}$$
$$= 179\cdot3 \text{ kJ}$$

THE RELATION BETWEEN THE SPECIFIC HEAT CAPACITIES OF A GAS c_p AND c_v

Let m kg of gas be receiving heat and expanding at constant pressure in a closed system from temperature T_1 to temperature T_2.

Since the heat required to raise the temperature of 1 kg of gas through 1 K at constant pressure is c_p,

Heat taken in $Q = mc_p(T_2 - T_1)$

When expanding, the gas does external work by forcing the piston outwards against a resisting force, i.e. work flows outwards from the system across its boundaries.

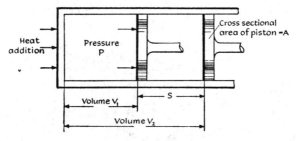

Fig. 12. Relation between Specific Heats of a Gas c_p and c_v

Let p be the constant pressure of the gas which is exerted on a piston of cross sectional area A, and let the piston move through a distance S. Heat is added to keep the pressure constant.

Constant force on piston $= pA$.

Now work done = Average force × distance moved
$$= pA \times S$$
$$= p(V_2 - V_1)$$

where V_1 and V_2 are the initial and final volumes of the gas.

$$\therefore \text{ Work done } W = p(V_2 - V_1)$$

Change of internal energy $(U_2 - U_1) = mc_v(T_2 - T_1)$

From the first law, for a closed system being conducted through a non-flow process:—

$$Q = W + (U_2 - U_1)$$
$$\therefore \ mc_p(T_2 - T_1) = p(V_2 - V_1) + mc_v(T_2 - T_1).$$

From the Characteristic Gas Equation

$$pV_2 = mRT_2 \text{ and } pV_1 = mRT_1$$
$$\therefore \ mc_p(T_2 - T_1) = mR\,(T_2 - T_1) + mc_v(T_2 - T_1)$$
$$\therefore \ c_p = R + c_v$$
$$\text{or } c_p - c_v = R$$

Example 8

'The volume of 0·18 kg of gas at a temperature of 15°C and pressure of 103 kN/m² occupies a volume of 0·15 m³. If c_v for the gas = 722 J/kgK, find

(a) The gas constant
(b) The molecular weight of the gas
(c) The specific heat at constant pressure
(d) The ratio of the specific heats.'

(a) $\dfrac{pV}{T} = mR$

$$\therefore \ R = \frac{pV}{mT}$$

$$= \frac{103 \times 10^3 \times 0·15}{0·18 \times 288} = \underline{298 \text{ J/kgK}}$$

(b) $R = \dfrac{8314}{M}$

$$\therefore \ M = \frac{8314}{298}$$
$$= \underline{27·9}$$

(c) $c_p - c_v = R$

$$\therefore\ c_p = 298 + 722$$
$$= 1020 \text{ J/kgK}$$

(d) $\gamma = \dfrac{c_p}{c_v} = \dfrac{1020}{722} = \underline{1 \cdot 412}.$

REVISION EXERCISES—CHAPTER 2

1. The molecular weight of a combustible gas is 13·12. Calculate the mass of 0·004 m³ of this gas at s.t.p. If this quantity of gas is contained in a gas-air mixture, the ratio of air to gas by volume being 10, calculate the mass of air in the mixture and the air-to-gas ratio by weight. Take the molecular weight of air as 28·9. Universal gas constant = 8314 J/kmolK.
   (Ans. 0·051 59 kg; 22·03 : 1)
2. A petrol engine has bore 100 mm, stroke 110 mm and compression ratio 5·5. When the crank is at outer dead centre the cylinder is full of gas mixture at a pressure of 100 kN/m² and temperature 35°C. Calculate the mass of the cylinder contents if M for the gas is 29·8. When the gas is compressed into the clearance volume, the pressure is found to be 1 MN/m². Determine the temperature at this point.
   (Ans. 0·001 23 kg, 287°C)
3. A cylindrical tank 0·75 m diam. and 1·6 m long stores gas at a pressure of 5 MN/m² and temperature 20°C. When some of the gas is tapped off, the pressure falls to 4 MN/m² and the temperature to 15°C. Taking M for the gas = 24, calculate the mass of gas taken out.
   (Ans. 6·47 kg)
4. A receiver which is used to measure the performance of an air compressor is a vertical drum 1·5 m diam. and 2 m high. At the start of the test the pressure and temperature of the air in the receiver are the same as that of the atmosphere. After 10 minutes running of the compressor, the receiver pressure is 700 kN/m² and the temperature 125°C. Determine the mass of air dealt with by the compressor in kg per minute, and express this in m³ of air per min at room conditions. Take atmospheric pressure and temperature as 750 mm of mercury, and 15°C respectively.
   R for air = 287 J/kgK.
   (Ans. 1·734 kg/min, 1·433 m³/min)
5. A petrol engine of bore 95 mm and stroke 125 mm has a clearance volume of 240 cm³. If the pressure and temperature of the gas

at the beginning of the compression stroke are respectively 101·3 kN/m² and 60°C, find the temperature at the end of compression when the pressure is 800 kN/m², and the change of internal energy during the stroke.

Take R for gas = 300 J/kgK and the ratio of specific heats $\gamma = 1·4$.

(Ans. 287·6°C, 195·2 J)

6. An air compressor takes in 0·5 kg of air at 15°C and compresses it to a temperature of 185°C. If 130 kJ of work are done on the air, determine the heat passing through the cylinder walls to the water jacket during the process.

Take $R = 287$ J/kgK, and $c_p = 1005$ J/kgK for air.

(Ans. 68·97 kJ)

7. 1 kg of air initially at 100 kN/m² and 20°C are compressed in a closed system through a volume ratio of 6 to 1 to a final pressure of 1·4 MN/m². Heat is then added at constant volume to increase the temperature to 550°C.

(a) Determine the change of internal energy during the compression process. ˙

(b) Calculate the amount of heat added at constant volume. Take c_v for air = 717 J/kgK.

(Ans. 280·2 kJ; 99·88 kJ)

8. Heat is added to a gas in a closed system whilst its volume increases from 28 dm³ to 100 dm³ at a constant pressure of 700 kN/m². If the molecular weight of the gas is 28 and its initial temperature is 50°C, determine (a) The mass of the gas; (b) The change of internal energy; (c) The external work done; (d) The heat flow, stating the direction in which it occurs.

Take $c_v = 717$ J/kgK; Universal gas constant = 8314 J/kmolK.

(Ans. 0·2044 kg, 121·6 kJ, 50·4 kJ, +172 kJ (intake))

3. Internal-combustion Engines Operation and Testing

We may divide heat engines into two main groups as follows:—

(a) Internal-combustion Engines, in which the energy is given to the working agent by burning fuel inside it. In this group are petrol, oil and gas engines, where the working agent is mainly air.

(b) External-combustion Engines, in which energy is released from the fuel in a separate furnace, and is transferred to the working agent across a separating wall. In this group are steam engines and steam turbines, where the working agent is steam.

In any internal-combustion engine the following requirements must be met:—

  (i) Fuel and air in the correct proportion must be supplied to the engine.

  (ii) The fuel and air must be compressed either before or after the mixing takes place.

  (iii) The compressed mixture must be ignited and the resulting expansion of the combustion products used to drive the engine mechanism.

  (iv) The exhausted combustion products must be cleared from the engine when their expansion is complete, in order to make way for a fresh charge.

Two methods are used to carry out these processes in a reciprocating I.C. engine, namely the Four-stroke Cycle, and the Two-stroke Cycle.

THE FOUR-STROKE CYCLE

This is so called because four strokes of the piston are required to complete the cycle, during which time the crankshaft turns through two revolutions. In petrol and gas engines, the mixing of the fuel and air takes place outside the cylinder, and the mixture is then drawn into the cylinder and compressed before being fired by an electric spark. In oil engines (commonly called diesel engines), however, only the air is drawn in and compressed by the piston, the oil being pumped into the cylinder when the compression is complete. In this way, the oil is fired by coming into contact with the hot air which raises the oil above its self-ignition temperature. A diesel engine needs no sparking plug or other separate ignition equipment.

Stroke 1—The Suction Stroke

This stroke begins with the piston at top dead centre. The crank is rotated by the energy from a flywheel, or by a starter motor if the engine is just being started, and the piston moves down the cylinder. With the exhaust valve closed and the inlet valve open, the downward movement of the piston causes suction in the cylinder which draws in a fresh charge through the inlet valve. This consists of

(a) Petrol-air mixture in the case of the petrol engine, or (b) Air only in the case of the diesel engine.

Stroke 2—The Compression Stroke

With both valves closed, the piston moves up the cylinder and compresses the charge into the clearance space. At, or near, the top of the stroke the charge is fired, either

(a) In the petrol engine by a spark which is made to occur at the points of a sparking plug which penetrates into the clearance space, or (b) In a diesel engine, by the oil being pumped into the clearance space in the form of a fine spray. This oil meets the air which has been much raised in temperature by being compressed, so that the oil heats up and fires spontaneously.

In either case, ignition causes heat energy to be released from the fuel by combustion, which results in a rise in pressure of the gases.

Stroke 3—The Expansion Stroke

Both valves remain closed, and the high pressure of the gases drives the piston out on its power stroke.

Stroke 4—The Exhaust Stroke

When the piston is near the bottom of the stroke, the exhaust valve opens whilst the inlet valve remains closed. The piston now moves up the cylinder again, driving before it the burnt-out gases

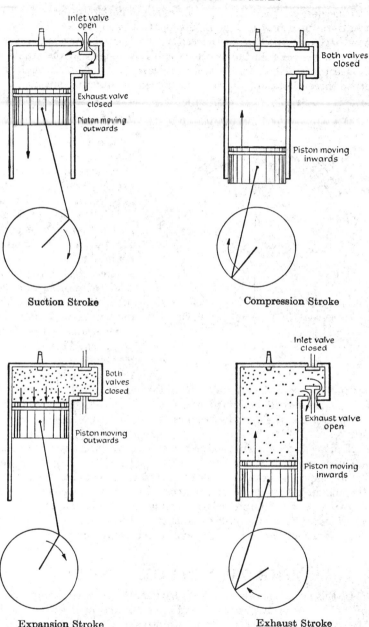

Fig. 13. Four-stroke Cycle

and discharging them through the exhaust port to prepare for the admission of a new charge on the next suction stroke. The cycle now begins again, and the procedure is repeated. As there is only one power stroke during two complete revolutions of the crank, a flywheel in which energy is stored to carry the engine through the other three strokes of the cycle is carried on the crankshaft.

The Valve-timing Diagram for the Four-stroke Cycle

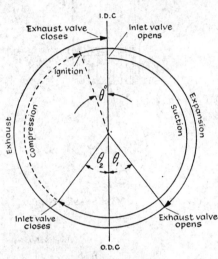

Fig. 14. Valve-timing Diagram

This is a most useful diagram which shows the correct crank positions for opening and closing the two valves in order to obtain the best performance from the engine. Fig. 14 is a typical valve-timing diagram for a four-stroke petrol engine. The inlet valve opens at inner dead centre, and remains open until well after the outer dead centre position is passed. This enables a greater charge weight per cycle to enter the cylinder by making use of the inertia of the inflowing gases. The closure of the inlet valve should be timed to coincide with the instant at which the pressure in the cylinder reaches atmospheric on the compression stroke.

Similarly, to obtain the best cleaning or 'scavenging' of the exhaust gases, it is usual to have the exhaust valve opening well before the crank reaches outer dead centre on the expansion stroke. In this way the speed of the gases through the exhaust valve is increased, and the improved scavenging more than makes up for the small loss of power from the expansion stroke. Closure of the exhaust valve is usually at or near the inner dead centre position.

THE TWO-STROKE CYCLE

In this cycle the operations of charging the cylinder, compression of the mixture or air, expansion of the gases and final scavenging of the cylinder are carried out in two strokes of the piston with one revolution of the crank. To achieve this, it is necessary to have a

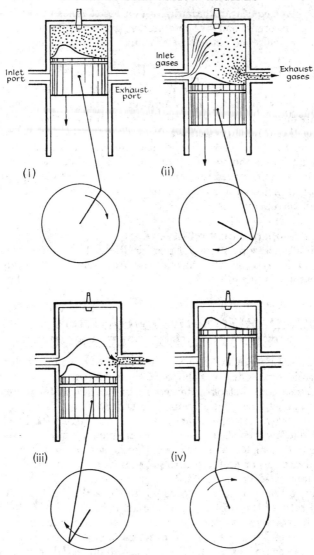

Fig. 15. Two-stroke Cycle

pump outside the cylinder in which the air is compressed so that it can force its way into the cylinder under pressure. The pump is usually part of the engine, and is driven by it, or the crankcase itself is designed as a compressor using the pumping action of the underside of the piston.

The valves consist of diametrically opposite holes in the cylinder walls at the crank end of the cylinder, and the piston is specially crowned to help the actions of charging and scavenging.

Stroke 1—Piston moving outwards

Starting with the cylinder full of fresh charge, release of heat energy occurs, either by spark ignition in the petrol engine or by spontaneous ignition in the diesel engine due to the injected oil meeting the hot compressed air. The expanding gases drive out the piston on the power stroke, and near the bottom of its stroke the piston uncovers the valve ports. The fresh charge is forced into the cylinder under pressure, and is deflected round the cylinder by the shaped piston head so that it helps to drive the exhaust gases out through the exhaust port.

Stroke 2—Piston moving inwards

Soon after leaving the outermost position, the piston covers both ports again and on this stroke the fresh charge is compressed preparatory to firing.

COMPARISON OF TWO-STROKE AND FOUR-STROKE CYCLE ENGINES

A two-stroke cycle engine has twice as many power strokes as a four-stroke cycle engine at the same engine speed. We should, therefore, expect a two-stroke engine to develop twice the power of a four-stroke engine of the same dimensions. This is nearly true, especially with slow-speed diesel engines, although with high-speed two-stroke engines the short time available to introduce the fresh charge and clear out the used gases leads to a diluted mixture and consequent loss of power. Further losses of power occur due to the power absorbed in driving the air pump and the reduction of the effective stroke due to the valve ports. Nevertheless, the power-to-weight ratio of the two-stroke cycle engine is much better than that of the four-stroke, and the two-stroke cycle is becoming increasingly used for diesel engines. In addition, the simple valve ports of the two-stroke engine which eliminate all the complicated valve mechanism of the four-stroke engine are a decided advantage.

There are, however, many more four-stroke engines in use than there are two-stroke engines, particularly in the petrol engine field. The thermal efficiency of a four-stroke engine is usually better, because the two-stroke engine has an increased specific fuel consumption (kg/kWh), owing to fuel losses through the exhaust port during the 'valve-open' period. The two-stroke cycle engine also tends to use more lubricating oil.

THE ENGINE INDICATOR

Before examining in more detail some of the engines which work on the four-stroke or two-stroke cycles, we must refer to a most useful instrument which gives a picture of the happenings in the engine cylinder throughout the cycle of operations. The engine indicator, which was invented by Watt for use on his steam engines, draws a graph of gas pressure against cylinder volume during a complete cycle.

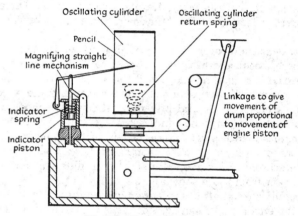

Fig. 16. Engine Indicator

In the simple indicator shown in Fig. 16, the pressure of the gases in the cylinder acts on a small piston and moves it against a spring. The displacement of this piston, which depends on the intensity of the gas pressure and the stiffness of the spring used, is magnified by a lever and recorded by a pencil on a paper-covered drum. The height of the line so recorded is thus a measure of the gas pressure to some scale.

At the same time, the base of the drum is connected through a mechanism to the engine piston, so that the drum rotates backwards and forwards as the piston moves backwards and forwards. The pencil is now recording cylinder pressure at each point throughout the cycle, and a graph results rather like Fig. 17 for a four-stroke cycle engine.

The scale of the graph can be determined from the spring number

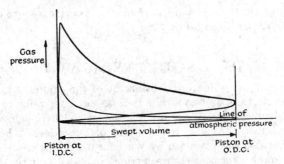

Fig. 17. Cylinder-pressure Graph

of the indicator spring (which takes into account the size of the indicator piston and the magnification of the lever mechanism) for the pressure ordinate, and from the engine dimensions for the volume abscissa.

Actual pencil indicators are a good deal better job than the one shown here, because great care is taken to design the magnifying lever mechanism so that the pencil moves, as nearly as possible, parallel to the indicator piston. Students should make a point of taking a careful look at, and preferably stripping, a modern pencil indicator. Also, these pencil indicators are not suitable for high-speed engines because, due to the inertia of the moving parts, the spring cannot recover quickly enough. However, successful high-speed indicators are made, although with different methods of measurement from those shown here.

A carefully taken indicator diagram gives a great deal of information to the engineer. Primarily, since it is a pressure-volume graph (and hence a force-distance graph since $pV = (p \times A) \times l$) it indicates the work being done on the engine piston, but it also gives valuable information about ignition timing, valve operation and the process of combustion.

THE PETROL ENGINE

Most petrol engines operate on the four-stroke cycle, and a typical indicator diagram is shown in Fig. 18. The air and petrol are mixed in the correct proportion in the carburettor and are drawn into the cylinder during the suction stroke $(a - b)$. The mixture is compressed during the compression stroke and the inlet valve is arranged to close at the point on the compression curve when the pressure in the cylinder reaches atmospheric pressure. Near the end of the compression stroke the charge is fired from a sparking plug, and as heat is released from the fuel the pressure in the combustion space rises

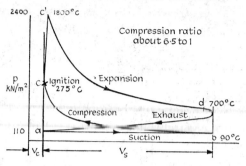

Fig. 18. Indicator Diagram

rapidly. The indicator diagram shows that this heat-release occurs whilst the volume of the gases remains approximately the same, i.e. heat addition at constant volume (c to c^1).

The high-pressure gases drive out the piston on the expansion stroke ($c^1 - d$), and upon the opening of the exhaust valve at d the exhaust gases are ejected to the atmosphere.

The provision of a correct air-fuel mixture and a precisely timed spark are of such importance that an introduction to the techniques used is now given.

The Carburettor

As we shall see from the work on the combustion of fuels, 1 kg of petrol needs about 15 kg of air for efficient burning. The purpose of the carburettor is to maintain the correct mixture of petrol and air to the engine, and to ensure that complete mixing takes place.

Fig. 19 shows a simple carburettor.

A petrol pump delivers petrol to the float chamber, the delivery being controlled by a tapered needle which ascends as the float rises. The petrol is led from the float chamber to a jet situated in the main air stream, and the air duct or choke tube is reduced in cross section at the point of entry of the petrol. The level of the exit from the jet is the same as that of the petrol in the float chamber, so that when the engine is not running, and the pressure in the choke tube is atmospheric, the jet is full of petrol. On the suction stroke of the piston, the pressure in the choke tube is reduced, causing the petrol to flow from the jet and mix with the air being drawn into the engine. The amount of mixture delivered to the engine is controlled by means of a butterfly throttle.

Unfortunately, a simple carburettor like this would not work very well in practice. Experiment shows that, due to friction and other

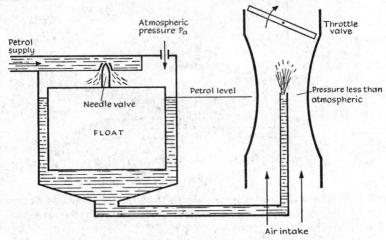

Fig. 19. Carburettor

effects, as the engine speed increases the proportion of fuel to air increases, i.e. the mixture becomes progressively richer when using a simple carburettor. In order that the carburettor shall give a constant mixture strength, a compensating jet is added (Fig. 20).

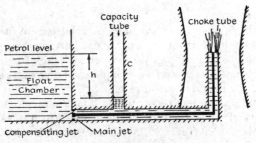

Fig. 20. Carburettor Jets

The main jet operates in the same way as in the simple carburettor, i.e. the mixture strength increases with engine speed. The compensating jet feeds a well (or 'capacity tube') C which is open to atmosphere. When the engine is stationary, the level of petrol in the capacity tube is the same as that in the float chamber. At a very small engine speed, however, the capacity tube empties, and the flow of the petrol through the compensating jet depends only on the size of the jet and the head of petrol above it (h) in the float chamber, both of which are constant. Thus above all but the lowest speeds, the flow of petrol through the compensating jet is constant. The air flow, however, increases with engine speed, so that the part of the mixture which is being made from the petrol of the compensating jet becomes steadily *weaker*. This balances the increasingly rich part of the mixture being made from the petrol of the main jet, so that by a suitable choice of main and compensating jet sizes the total mixture strength may be kept reasonably constant.

Modern carburettors are finely designed units offering different methods of compensation as well as additional arrangements for starting and idling, and in recent years there has been a tendency towards direct petrol injection.

The Ignition Circuit

The compressed air-and-petrol mixture must be fired at the correct instant so that the resulting rise in pressure acts on the piston

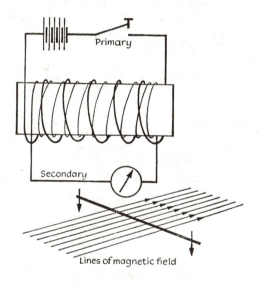

Fig. 21. Induction Coil

when it is at the top of the cylinder, and the expanding gases force the piston out on the power stroke. A high voltage is required to jump the gap of a sparking plug and give a spark of sufficient energy to ignite the mixture, and this is produced by an induction coil.

The induction coil consists of two separate insulated windings on an iron core, the PRIMARY being a small number of turns of thick wire, and the SECONDARY a large number of turns of thin wire.

45

When an electric current is flowing through the primary circuit, a magnetic field is set up along the iron core and through the secondary windings.

Experiment shows that when a wire is made to move across a magnetic field, an electric current starts to flow in the wire. The same effect can be obtained by holding the wire still and switching the magnetic field on or off. Thus, in the induction coil, if the flow of electricity in the primary winding, which produces the magnetic field, is switched on or off, a flow of electricity is induced in the coils of the secondary winding, which lie across the field. The voltage of the induced current depends on the relative number of turns in the coils, and in a car engine an interruption in the primary circuit using a battery of 12 volts, induces about 11 000 volts in the secondary circuit.

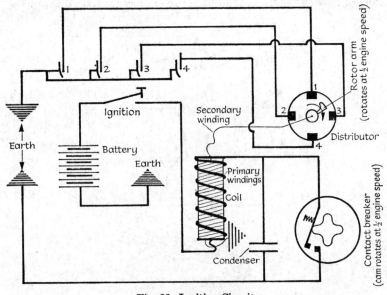

Fig. 22. Ignition Circuit

The ignition circuit for a four-cylinder petrol-engine working on the four-stroke cycle is shown in Fig. 22.

The primary coil is energized from a battery, and the high-voltage secondary current is induced at each break in the flow of the primary current. The break occurs at the contact breaker which is actuated by a cam driven at half engine speed. The cam has as many lobes as there are cylinders, and the high voltage current from the secondary windings of the induction coil is directed to each sparking plug in turn by a distributor which rotates with the cam.

It takes some time after the occurrence of the spark for the fuel to ignite and release its heat, so that arrangements must be made for ignition to occur before top dead centre so that the pressure begins to rise just as the piston reaches the top of its stroke. Such an operation is called 'advancing the ignition'. In variable speed engines the angle turned through by the crank during the ignition delay period varies with engine speed, and it is, therefore, necessary to have a device to increase the angle of advance as the speed increases. This is usually carried out automatically by a small centrifugal governor which alters the angular position of the cam operating the contact breaker.

THE DIESEL ENGINE

The essential difference between a diesel engine and a petrol engine lies in the method of ignition. In the diesel engine, air alone is compressed during the compression stroke to a temperature above the ignition temperature of the oil fuel which is then forced into the cylinder under pressure. Due to the heavy nature of the oil fuel, steps must be taken to ensure adequate mixing of the fuel and air. Diesel engines may be divided into two main groups, those running from, say, 75 to 250 rev/min called slow-speed diesels, and those from 250 to 1200 rev/min called high-speed diesels.

Slow-speed Diesels

This type of diesel engine has been used mainly for marine engines and works on the two-stroke cycle. The fuel is forced into the cylinder by a blast of compressed air, and the rate of injection is such as to

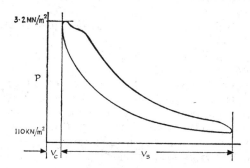

Fig. 23. Slow-speed Diesel Diagram

maintain the gases at approximately constant pressure during the combustion period. Fig. 23 shows a representative indicator diagram.

47

High-speed Diesels

By far the greater number of diesel engines operate at such a speed that there is insufficient time available to heat and burn the injected oil efficiently if the injection is delayed until the air is compressed sufficiently. The injection of the oil is therefore begun before the end of the compression stroke; and before the air is hot enough to ignite it immediately it enters the cylinder. There follows a *delay period* during which the injected oil accumulates, so that when the compression of the air has proceeded to ignition temperature a sudden large rise in pressure follows the burning of the accumulated oil. As further injection continues, the pressure stays approximately constant, so that the combustion is partly at constant volume and partly at constant pressure.

The indicator diagram in Fig. 24 represents the process for a four-stroke cycle engine. The maximum pressure is much greater than

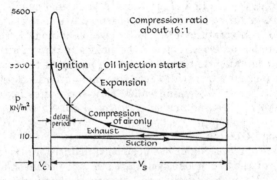

Fig. 24. High-speed Diesel Diagram

that in the petrol engine, so that a diesel engine is more heavily built to withstand it.

The Injection System

High-speed diesel engines need an injection system consisting of a nozzle which penetrates into the combustion space and from which the oil emerges in a fine spray, and a pumping device to supply the oil under pressure to the nozzle. The C.A.V. (Bosch) pump is widely used and enables the quantity of oil being pumped to meet the requirements of varying load or speed.

Each cylinder of an engine is supplied by a separate pump and nozzle. The plunger of the pump which is operated by a cam has a constant stroke, and can be rotated in the plunger cylinder so as to control the amount of fuel pumped to the nozzle. The vertical

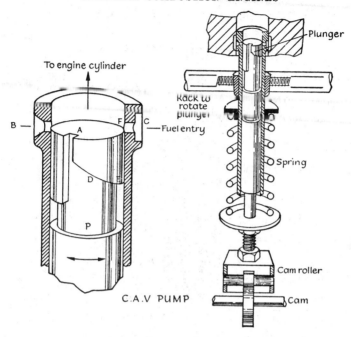

Fig. 25. Injection System

groove A leads into the helical groove D. On the upstroke of the plunger, when ports B and C are covered, the fuel above the pump and in the undercut section D is under pressure, and delivery to the engine via the nozzle takes place until the edge E uncovers the port F. Pressure is then relieved and the delivery stops, more fuel being drawn in at the bottom of the downstroke when suction is created. The plunger P may be rotated from the throttle control so that the length FE may be varied which varies the length of the effective stroke and hence the quantity of fuel delivered to the engine.

The nozzle is required to spray the fuel into the combustion chamber so that it is adequately atomized or broken up. In order to achieve this, injection takes place through very fine holes in the nozzle at a pressure of about 17.5 MN/m^2. Fig. 26 shows the principle of a Gardner-type nozzle. Injection begins when the valve is lifted from its seating, against the force of a pre-loaded spring, by the oil pressure on the annular area between the two diameters of the valve stem. Because of the small diameter of holes used in the nozzle, very careful filtering of the fuel is essential.

49

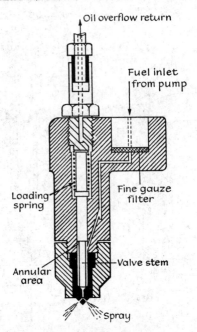

Fig. 26. Gardner-type Nozzle

MEASUREMENT OF BRAKE POWER

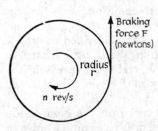

Fig. 27. Measurement of Brake Power

Work is done when a resisting force F is overcome through a certain distance S, the amount of work being measured as the product FS. Power is the rate of doing work.

If a braking force F is applied at the rim of a wheel of radius r, the product Fr is called the resisting torque T. If the wheel is kept turning at N rev/s against this braking action, then

Work done per second = Resisting Force F × Distance through which the force is overcome per second.

$$= F \times 2\pi r \times N$$

50

(for 1 rev, distance moved against F is the circumference $2\pi r$)

i.e. Work done per second $= 2\pi N \times Fr$
$= 2\pi NT$ (Since $F \times r =$ torque T)
$= T\omega$ (watts)

i.e. b p $= T\omega$ watts, where $T =$ resisting torque (Nm)
$\omega =$ angular velocity (radians/second)

Any method for measuring the brake horsepower of an engine involves the application of a torque which resists the motion of the crankshaft.

One method is to use the simple rope brake, illustrated in Fig. 28.

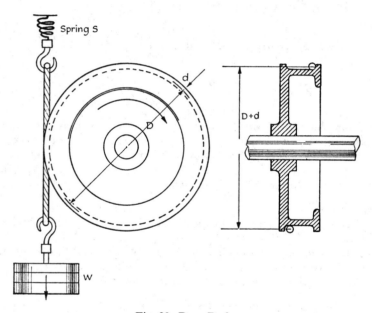

Fig. 28. Rope Brake

A rope makes one complete turn round the rim of a flywheel keyed to the engine crankshaft. The rope carries a dead load on one end and is hooked to a spring balance at the other, the direction of the rotation being against the pull of the dead load. The engine is started with the load off, and increasing load may be applied by adding weights to the dead-load hanger. At any steady load, with the engine running at a constant speed:—

51

let W = dead load on brake (kgf)

S = Spring balance reading (kgf)

D = Diameter of brake wheel (m)

d = diameter of rope (m)

Consider the load to be acting along the centre line of the rope:—

Resisting torque due to dead load $= 9 \cdot 81 \ W\left(\dfrac{D+d}{2}\right)$ Nm

(Note that the force due to the weight of a mass of 1 kilogramme is $9 \cdot 81$ newtons.)

Notice that the torque due to the spring load S acts *in the same direction* as the direction of rotation of the wheel, so that it reduces the resisting torque.

Torque due to spring pull $= 9 \cdot 81 \ S\left(\dfrac{D+d}{2}\right)$ Nm

∴ Effective braking torque $T = 9 \cdot 81 \ W\left(\dfrac{D+d}{2}\right) - 9 \cdot 81 \ S\left(\dfrac{D+d}{2}\right)$

$$T = 9 \cdot 81 \ (W - S)\left(\dfrac{D+d}{2}\right) \ \text{Nm}$$

And b.p. $= T\omega$ watts

From the First Law of Thermodynamics, we see that the work done cannot just disappear, but that the energy will be converted into another form. In this case heat will be generated, and so arrangements must be made to keep the brake cool by a flow of water inside the rim of the brake wheel.

Example 9

'In a test on a single-cylinder gas engine using a simple rope brake, the following readings were taken:—

Dead load 29 kgf, spring balance reading 4 kgf, speed 284 rev/min, diameter of brake wheel $1 \cdot 05$ m, diameter of rope = 20 mm.

Calculate the b.p. being developed by the engine.'

Braking torque $= 9 \cdot 81 \ (W - S)\left(\dfrac{D+d}{2}\right)$ Nm $\quad D + d = 1 \cdot 05 + 0 \cdot 020$

$\qquad\qquad\qquad\qquad\qquad\qquad\qquad\qquad\qquad = 1 \cdot 07$ m

$\qquad = 9 \cdot 81 (29 - 4)\left(\dfrac{1 \cdot 07}{2}\right)$

$\qquad = 131 \cdot 2$ Nm

b.p. $= T\omega$

$\qquad = \dfrac{131 \cdot 2 \times 2\pi \times 284}{60 \times 1000}$

$\qquad = \underline{3 \cdot 902 \ \text{kW}}$

52

Simple rope brakes can be safely used only on low-speed engines which have their speed kept reasonably constant by a governor. For high-speed engines, a widely used brake is the Heenan and Froude Hydraulic Dynamometer. Consider the engine to drive a wheel, made up of a series of buckets, in an enclosed chamber full of water. Fig. 29 (a) shows the method diagramatically.

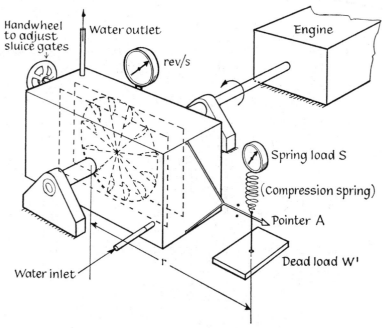

Fig. 29 (a). Heenan and Froude Hydraulic Dynamometer

The casing is freely mounted, so that when the wheel begins to turn and pushes against the water, the casing would turn with it were it not prevented from doing so.

The casing is prevented from rotating by the force of a spring S and a dead load W, applied to an arm which forms part of the casing. When the casing is level, shown by a pointer A, then the torque or turning effect of the engine, which is being transmitted to the casing through the bucket wheel, is being balanced by the equal and opposite resisting torque due to the load $(W_1 + S)$ being applied through a radius r

i.e. Resisting torque $= 9\cdot81\ (W_1 + S)\ r$ Nm

$\therefore$ b.p. $= T\omega = 9\cdot81\ (W_1 + S)\ r\omega$ watts

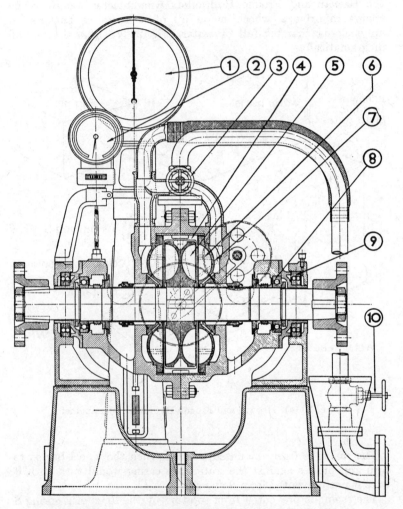

Fig. 29 (b). Typical Cross Section Through Casing of Froude
Dynamometer, Type D.P.Y.

1. Spring Balance.
2. Tachometer.
3. Water Outlet Valve.
4. Sluice Gates for Load Control.
5. Rotor.

6. Water Inlet Holes in Vanes.
7. Casing Liners.
8. Shaft Bearing.
9. Casing Trunnion Bearing.
10. Water Inlet Valve.

In the actual dynamometer, a cross section of which is shown in Fig. 29 (b), the rotating wheel consists of a series of half-elliptical pockets, running between an identical set of pockets in the stationary casing. The motion of the water set up by the wheel is transferred to the pockets in the casing and tends to turn the casing with the shaft. Between the wheel pockets and the casing pockets are thin plates which can be moved using an external control, and so blank-off a number of working pockets. This gives control over engine speed for a fixed throttle setting.

In the expression b.p. $= 9\cdot81\ (W_1 + S)\ r\ 2\pi n,$

since r is a fixed length for a given machine, we may write

$$9\cdot81 \times 2\pi r = \text{constant}\quad\text{and Load } W = (W_1 + S)$$

i.e. b.p. $= Wn \times \text{Constant}$

MEASUREMENT OF INDICATED POWER

An indicator diagram is a pressure-volume graph of the conditions in the cylinder throughout a complete cycle. A typical diagram from a gas engine is shown in Fig. 30 (a), p. 56.

We may write $p \times V = p \times A \times l$ [Since Volume = Area × length]
$= \text{Force} \times \text{distance}$ [Since pressure × Area = Force]
$= \text{Work}$

Thus: The area under an expansion curve, i.e. when volume is increasing, represents work done *by* the gases, i.e. positive work (Fig. 30 (b) and (d)). The area under a compression curve, i.e. when volume is decreasing, represents work done *on* the gases, i.e. negative work (Fig. 30 (c) and (e)).

By adding algebraically the areas for each operation throughout the cycle, the diagram may be seen to consist of two enclosed areas, the larger of which represents the work done by the engine each cycle. The smaller enclosed area is called the PUMPING LOOP and represents a loss of work from the engine resulting from the necessity to clean and recharge the cylinder. We shall ignore, for the moment, the pumping loop, because it is usually too small to measure on a normal diagram.

The vertical scale of the diagram depends on the spring used in the indicator, and is given as the spring number with the units of pressure per unit length, e.g. N/m^2 per mm; kN/m^2 per metre of MN/m^2 per metre. You will need to check carefully the springs used in the indicator you used, for until the introduction of S I units springs were marked with numbers representing lbf/in^2 per in and conversion will be necessary.

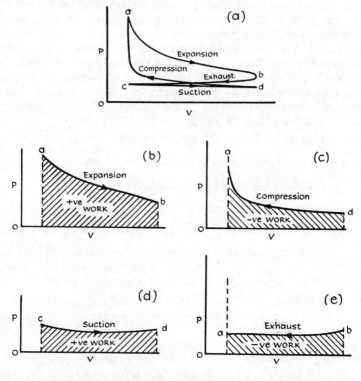

Fig. 30. Measurement of Indicated Horsepower

MEAN EFFECTIVE PRESSURE

The average net pressure which, acting on the piston area for one stroke, does the same work as that represented by the indicator diagram is known as the 'Mean effective pressure, P_m'. The enclosed area of the diagram is irregular in shape, but a rectangle of equal area and having the same base length would have a height equal to the mean effective pressure.

Hence to determine the mean effective pressure from a given indicator diagram, measure the closed area using a planimeter, and then

$$\text{Mean effective pressure } P_m \, (\text{N/m}^2) = \frac{\text{Area of diagram (mm}^2)}{\text{Base length of diagram (mm)}} \times \frac{\text{Spring No.}}{(\text{N/m}^2 \text{ per mm})}$$

Now, the indicated power can be calculated using the mean effective pressure as follows:—

56

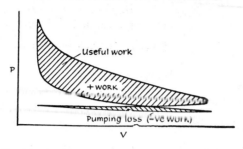

Fig. 31. Graph showing Work Done and Pumping Loss

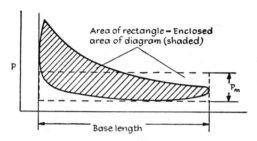

Fig. 32. Graph of Mean Effective Pressure

i.p. = Work done per second inside the cylinder (Nm/s = J/s = W)
    = Work done per cycle (Nm = J) × Number of cycles/second (W)

Now work done per cycle = Area of indicator diagram
$$= P_m \times V_s.$$
$$= P_m \times (L \times A)$$
$$\therefore \text{ i.p.} = P_m LAn \text{ (Watts)}$$
where P_m = Mean effective pressure (N/m²)
    L = Stroke (m)
    A = Piston area (m²)
    n = Number of working cycles per second.

Notice that when mean effective pressure P_m (N/m²) is multiplied by piston area (m²), the result is a force in Newtons. Hence the stroke L must be given in metres to produce the units of work, Nm = Joules.

For a four-stroke cycle engine, the number of cycles per second is half of the engine speed (i.e. one power stroke every two revolutions).

For a two-stroke cycle engine, n = the engine speed.

57

For a gas engine, where the firing is controlled by a hit-and-miss governor, it is necessary to count the number of firing strokes. If no misses occur, then on the four-stroke cycle $n = \dfrac{\text{Engine speed}}{2}$;

otherwise, number of firing strokes $n = \dfrac{\text{Engine speed}}{2} - \text{Number of}$ misses/second.

PUMPING LOOP

We should now take account of the loss of power as shown in the pumping loop. To obtain an accurate value of the mean effective pressure during the pumping loop, a weak spring is fitted to the indicator which shows the *pumping loop* to a larger scale.

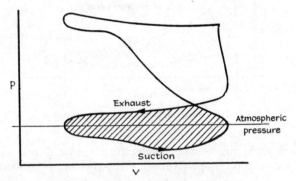

Fig. 33. Graph showing Pumping Loop

The mean effective pressure of the pumping loop is found by one of the methods described, so that the pumping horsepower taken from the engine is given by:—

$$\text{Pumping power} = P_p L A n \text{ watts}$$
$$P_p = \text{mean effective pressure of pumping loop (N/m}^2\text{)}$$

I.P. of Multicylinder Engines

For engines having more than one cylinder, the most accurate method is to measure the i.p. of each cylinder separately. The i.p. of the engine is the sum of the i.p.'s of the separate cylinders. A good approximate result, however, can be obtained by finding the i.p. of one cylinder, whence

$$\underset{\text{engine}}{\text{i.p.}} = \underset{\text{one cylinder}}{\text{i.p.}} \times \text{Number of cylinders.}$$

A useful practical method of finding the i.p. of one cylinder of a multicylinder petrol engine without the use of a high-speed indicator is known as the MORSE TEST. Suppose we have a four-cylinder petrol engine on a testbed with a Froude dynamometer to measure b.p. At any given speed

$$\frac{\text{i.p.}}{\text{4 cyln.}} = \frac{-\text{b.p.}}{\text{4 cyls.}} + \frac{\text{friction power}}{\text{4 cyls.}} \quad \dots\dots[1]$$

If the sparking plug lead to one cylinder is shorted, the engine speed will fall owing to the reduced power of the engine. The load on the brake may be reduced so that the engine speed increases again to the original speed selected for the test. Notice that now the engine is developing power in three cylinders only whereas the friction in all cylinders remains the same.

$$\text{Hence} \quad \frac{\text{i.p.}}{\text{3 cyls.}} = \frac{\text{b.p.}}{\text{3 cyls.}} + \frac{\text{friction power}}{\text{4 cyls.}} \quad \dots\dots[2]$$

Subtracting equation [2] from equation [1]:—

$$\frac{\text{i.p.}}{\text{4 cyls.}} - \frac{\text{i.p.}}{\text{3 cyls.}} = \frac{\text{b.p.}}{\text{4 cyls.}} - \frac{\text{b.p.}}{\text{3 cyls.}}$$

$$\text{But} \quad \frac{\text{i.p.}}{\text{4 cyls.}} - \frac{\text{i.p.}}{\text{3 cyls.}} \quad \text{is the i.p. of the cylinder that}$$

was cut out, and hence this may be calculated as the difference in readings of b.p. measured when all cylinders were firing and when one was cut out.

Example 10

'A four-cylinder petrol engine at 1200 rev/min gave 25·3 kW b.p. When one cylinder was cut out the b.p. decreased to 17·6 kW. Estimate the i.p. of the engine.'

$$\text{i.p.} = 4(25\cdot3 - 17\cdot6)$$
$$= 30\cdot8 \text{ kW}$$

Example 11

'In a test on a single-cylinder oil engine operating on the four-stroke cycle and fitted with a simple rope brake, the following readings were taken:—

Brake wheel diam. 0·7 m, speed 450 rev/min, rope diam. 20 mm, deadweight on rope 21 kgf, spring balance reading 3·4 kgf, area of indicator diagram = 404 mm², length of indicator diagram 65 mm, spring No. 140 kN/m² per mm, bore 100 mm, stroke 150 mm, engine used 0·75 kg/h of oil of calorific value 45 MJ/kg. Calculate the b.p., i.p., mechanical efficiency and indicated thermal efficiency of the engine.'

$$\text{b.p.} = T\omega \qquad T = 9\cdot81(21 - 3\cdot4)\left(\frac{700 + 20}{10^3 \times 2}\right)$$

$$= 62\cdot16 \text{ Nm}$$

$$\therefore \text{b.p.} = \frac{2\pi \times 450 \times 62\cdot16}{60}$$

$$= 2\cdot936 \text{ kW}$$

$$\text{i.h.p.} = P_m LAn \qquad P_m = \frac{\text{Area of diagram}}{\text{Length of diagram}} \times \text{Spring No}$$

$$= \frac{404}{65} \times 140 \times 10^3$$

$$= 870\cdot2 \text{ kN/m}^2$$

$$\text{i.p.} = 870\cdot2 \times 10^3 \times \frac{\pi}{4} \times (0\cdot1)^2 \times \frac{0\cdot15 \times 450}{60 \times 2}$$

$$= 3\cdot845 \text{ kW}$$

$$\text{Mechanical efficiency } \eta_m = \frac{\text{b.p.}}{\text{i.p.}} = \frac{2\cdot936}{3\cdot845} = 0\cdot764 \text{ or } 76\cdot4\%$$

$$\text{Indicated thermal efficiency} = \frac{\text{indicated output}}{\text{heat supplied per second}}$$

$$= \frac{3845 \times 3600}{0\cdot75 \times 45 \times 10^6}$$

$$= 0\cdot4102 \text{ or } 41\cdot02\%$$

Example 12

'During a test on a single-cylinder gas engine of 250 mm bore and 400 mm stroke working on the four-stroke cycle, the following readings were taken:—

Speed 240 rev/min, misses/minute = 10, area of indicator diagram = 520 mm², spring No. 0·1 MN/m² per mm, length of indicator diagram = 75 mm, effective brake load = 78 kgf, brake wheel diameter = 1·0 m, rope diameter = 30 mm. Gas used as measured at the meter = 50 m³/hour at a temperature of 20°C and pressure 90 mm water when barometer reading = 745 mm mercury. Take calorific value of gas = 16·5 MJ/m³ at 0°C and 760 mm. Calculate the b.p., i.p and thermal efficiency of the engine.'

$$\text{b.p.} = T\omega \qquad\qquad T = 9\cdot81 \times 78 \times \frac{1\cdot03}{2}$$

$$\therefore \text{b.p.} = \frac{2\pi \times 240 \times 9\cdot81 \times 78}{60} \times \frac{1\cdot03}{2} = 9\cdot91 \text{ kW}$$

$$\text{i.p.} = P_m L A n \qquad\qquad P_m = \frac{520 \times 10^5}{75} = 693\cdot0 \text{ kN/m}^2$$

$$\text{Firing Strokes} = \frac{240}{2} - \text{misses} = 120 - 10 = 110$$

$$\therefore \text{i.p.} = 693\cdot3 \times 10^3 \times \frac{\pi}{4} \times (0\cdot25)^2 \times 0\cdot40 \times \frac{110}{60} = 24\,960 \text{ W} = 24\cdot96 \text{ kW}$$

$$\text{Gas used at meter} = \frac{50}{3600} \text{ m}^3/\text{s}$$

$$\text{Pressure of gas used} = \left(745 + \frac{90}{13\cdot6}\right) \times 133\cdot3 = 100\cdot2 \text{ kN/m}^2$$

$$\text{Temperature of gas used} = 293 \text{ K}$$

Since the c.v. of the gas is given at 0°C and 760 mm of Hg (i.e. s.t.p.), we must convert the volume of gas under the measured conditions to what would have been used under s.t.p. conditions.

$$\frac{p_1 V_1}{T_1} = \frac{p_n V_n}{T_n}$$

$$\therefore V_n = \frac{p_1 V_1}{T_1} \times \frac{T_n}{p_n}$$

$$= \frac{100\cdot2}{293} \times \frac{50}{3600} \times \frac{273}{101\cdot3}$$

$$= 0\cdot012\,79 \text{ m}^3/\text{s}$$

$$\therefore \text{indicated thermal efficiency} = \frac{\text{i.p.}}{\underset{\text{used at s.t.p.}}{\text{m}^3 \text{ of gas}} \times \underset{\text{at s.t.p.}}{\text{c.v. of gas}}}$$

$$\therefore \text{indicated } \eta_t$$

$$= \frac{24\cdot96 \times 10^3}{0\cdot012\,79 \times 16\cdot5 \times 10^6}$$

$$= 0\cdot1182 \text{ or } 11\cdot82\%$$

61

Energy balance account for an I.C. engine regarded as operating in a closed system

An internal combustion engine operates as an open system undergoing a reasonably steady-flow process in which a mass of fuel and air enter the system and an equal mass of exhaust gas leaves the system. It is useful to have an account which shows how the energy entering the system has been distributed, but to determine such an account for an open system is complicated because of the need to determine such items as stored energy and flow work at the entry and exit points of the boundary. A rough approximation can be obtained, however, by regarding the engine as a closed system operating a cycle, in which case the First Law of Thermodynamics states that the heat flow across the boundary is equal to the work flow across the boundary (see page 20).

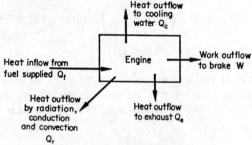

Fig. 34. Energy Balance Account

Such an arrangement is shown in Fig. 34, and applying the First Law:—

$$Q_f - Q_r - Q_c - Q_e = W$$
$$\text{or } Q_f = W + Q_r + Q_c + Q_e$$

This statement may be shown in the form of a balance sheet (often called a Heat Balance Account), the left-hand side showing the heat energy supplied and the right-hand side showing its distribution. The balance may be based on 1 kg of fuel, or on a basis of time, and the items may be expressed as heat quantities and as percentages of energy supplied. For I.C. engines, the table is usually as follows:—

(a) Energy supplied per second Q_f
     = Mass of fuel used/second × calorific value joules (watts)

(b) Energy distributed per second
     (i) Work outflow W = b.p. (watts)
     (ii) Heat flow to cooling water Q_c
          = $mc(t_{\text{out}} - t_{\text{in}})$ (watts)

62

where
$$m = \text{flow of cooling water kg/s}$$
$$c = \text{specific heat capacity}$$
$$t_{out} = \text{leaving temperature of cooling water}$$
$$t_{in} = \text{entry temperature of cooling water}$$

(iii) Heat flow to exhaust and surroundings $Q_e + Q_r$. It is difficult to measure the heat flow to the exhaust gases and to the surrounding air. In this simple form of presentation of a balance sheet, these are taken together as a joint item and are calculated as the difference between the energy already accounted for to work and cooling and the energy input.

Energy supplied/s	Joules	%	Energy distributed/s	Joules	%
kg fuel/s × c.v.			1. Work outflow		
			2. Heat flow to cooling water		
			3. Heat flow to exhaust and surroundings		
Total		100	*Total*		100

An addition to the balance sheet which is often asked for is the 'heat to friction'. This has already been referred to on page 13 as the reduction of power between the cylinder and the output shaft of an engine due to friction between the moving parts. Such friction converts some of the work developed at the cylinder head back again into heat, so that the work flowing from the engine is correspondingly reduced. Thus:—

$$\text{heat to friction} = (\text{i.p.} - \text{b.p.}) \text{ watts}$$

This item cannot, of course, be entered as a separate item in the account for it is already included as heat flowing from the engine partly in the cooling water, partly in the exhaust gases and partly in radiation to the surroundings.

Example 13

'The following results were obtained during the trial of a four-stroke national oil engine of cylinder bore 200 mm and stroke 400 mm:—

Effective brake wheel diameter = 1·6 m, speed 258 rev/min, effective brake load = 47 kgf, area of indicator diagram = 320 mm², spring No. 110 kN/m² per mm, length of diagram = 65 mm, fuel used/h $= 3 \cdot 2$ kg/h, c.v. of fuel $= 45$ MJ/kg, cooling water supplied $= 2 \cdot 3$ kg/min, rise in temperature of cooling water $= 38°C$. Calculate (a) Mechanical efficiency; (b) Indicated thermal efficiency, and (c) Draw up an energy balance account on a basis of 1 minute.'

$$\text{Brake torque } T = 9 \cdot 81 \times 47 \times \frac{1 \cdot 6}{2} = 368 \cdot 9 \text{ Nm}$$

$$\text{b.p.} = T\omega = 368 \cdot 9 \times 2\pi \times \frac{258}{60} = 9 \cdot 965 \text{ kW}$$

$$P_m = \frac{320 \times 110 \times 10^3}{65} = 0 \cdot 5416 \text{ MN/m}^2$$

$$\therefore \text{i.p.} = P_m LAn = 0 \cdot 5416 \times 10^6 \times \frac{\pi}{4} \times \frac{0 \cdot 2^2 \times 0 \cdot 4 \times 258}{2 \times 60}$$

$$= 14 \cdot 63 \text{ kW}$$

$$\therefore \text{ Mechanical efficiency} = \frac{\text{b.p.}}{\text{i.p.}} = \frac{9 \cdot 965}{14 \cdot 63} = 0 \cdot 681 \text{ or } 68 \cdot 1\%$$

$$\text{Indicated } \eta_t = \frac{\text{i.p.}}{\text{kg of fuel/s} \times \text{c.v.}} = \frac{14 \cdot 63 \times 10^3 \times 3600}{3 \cdot 2 \times 45 \times 10^6}$$

$$= 0 \cdot 3659 \text{ or } 36 \cdot 59\%$$

Energy input/min	J	%	Energy distribution/min	J	%
kg fuel/s × c.v. $= \frac{3 \cdot 2}{60} \times 45 \times 10^6$	$2 \cdot 4 \times 10^6$	100	1. To useful work $=$ b.p. × 60	$597 \cdot 9 \times 10^3$	24·92
			2. To cooling water $= m_c(t_{out} - t_{in})$ $= 2 \cdot 3 \times 4187 \times 38$	$365 \cdot 9 \times 10^3$	15·52
			3. To exhaust and radiation (by difference) $= [2400 - (597 \cdot 9 + 365 \cdot 9)]10^3$	$1436 \cdot 2 \times 10^3$	59·83
Total	2400×10^3	100	Total	2400×10^3	100

REVISION EXERCISES—CHAPTER 3

1. The diameter of the brake wheel fitted to a single-cylinder four-stroke oil engine on test is $0 \cdot 6$ m and the rope diameter is 25 mm. The dead-load is 20 kgf when the spring balance reading is 3 kg and the engine is running at 450 rev/min. From indicator cards the average m.e.p. $= 700$ kN/m². Fuel consumption is $0 \cdot 3$ kg/kWh (brake) of oil of c.v. 40 MJ/kg. Engine bore $= 100$ mm, stroke $= 150$ mm.

    Calculate the b.p., i.p., mechanical efficiency and indicated thermal efficiency at this load.

    (Ans. $2 \cdot 456$ kW, $3 \cdot 093$ kW, $79 \cdot 4\%$, $37 \cdot 78\%$)

2. A four-cylinder diesel engine working on the four-stroke cycle has cylinder diameter 200 mm and stroke 300 mm. On full load, the area of the indicator diagram taken from No. 1 cylinder is 480 mm², and the length of the diagram 65 mm. Indicator spring No. 80 kW/m² per mm. The effective diameter of the brake wheel is 1·8 m, and the applied load at the rim is 150 kgf. Speed 200 rev/min. Calculate the b.p., i.p. and mechanical efficiency of the engine.

(Ans. 27·74 kW, 37·12 kW, 74·7%)

3. A gas engine working on the four-stroke cycle is fitted with a hit-and-miss governor. On test, with an effective brake wheel diameter of 1·7 m, the load was 24 kgf at a speed of 306 rev/min. Area of indicator card = 450 mm², and length of diagram = 60 mm. Spring No. 80 kN/m² per mm. The gas consumption was 15 m³/h measured at 120 mm water gauge and 20°C when the barometer read 750 mm mercury; there are no 'misses' and the swept volume is 5×10^{-3} m³. Given c.v. of gas = 18 MJ/m³ at 0°C and 760 mm of Hg, calculate

(a) mechanical efficiency; (b) gas consumption in m³ at s.t.p./kWh (brake); (c) brake thermal efficiency.

(Ans. 83·83%, 2·176, 9·19%)

4. A diesel engine develops 12 kW at 360 rev/min, being supplied with oil of calorific value 45 MJ/kg. The specific fuel consumption is 0·27 kg/kWh (brake basis). One-third of the heat supplied is carried away by the cooling water which rises by 30°C in its passage through the engine. Calculate the brake thermal efficiency of the engine and determine the quantity of cooling water used in dm³ per minute.

(Ans. 29·63%, 6·45)

5. An engine on test has a specific fuel consumption of 0·35 kg/kWh (brake basis) and uses 0·068 kg/min of fuel of calorific value 43 MJ/kg. The flow of jacket cooling water is 300 kg per hour with its temperature rising from inlet 15°C to outlet 51°C. Draw up a simple heat balance account for the engine, and state the heat equivalent of the friction work given that indicated power is 14·3 kW.

(Ans. input = 2·924 MJ, W = 0·7 MJ, Q_c = 0·754 MJ, Q_{e+r} = 1·47 MJ, friction = 2·64 kW)

6. A single-cylinder gas engine working on the four-stroke cycle develops 6 kW at full load whilst using 90 dm³/min of town gas calorific value 18 MJ/m³. Flow of cooling water is 4·8 kg per minute with a temperature rise of 21°C. Determine the full-load brake thermal efficiency, and draw up a simple energy

balance account for the engine in MJ/min and in percentages.
(Ans. 22·2%, input 1·62, $W = 0·36$, 22·2%; $Q_c = 0·422$, 26·1%;
$Q_{e+r} = 0·838$, 51·7%)

7. A six-cylinder four-stroke petrol engine is required to produce 76 kW at 2000 rev/min when the mean effective pressure in each cylinder is 700 kN/m². If the design is to be that of a 'square-cylinder' engine, i.e. stroke = bore, calculate the required cylinder dimensions assuming a mechanical efficiency of 80%.

(Ans. $D = L = 120$ mm)

8. The following data were obtained during a test on a four-stroke oil engine having a bore and stroke of 300 mm and 450 mm respectively.

Speed = 200 rev/min, i.m.e.p. = 600 kN/m², fuel consumption = 8·8 kg/h, calorific value of fuel = 44 MJ/kg, net brake load = 190 kgf, effective diameter of brake = 1·25 m, cooling water flow 10·5 kg/min, mean temperature rise of cooling water = 20 degC.

Estimate i.p., b.p., indicated and brake thermal efficiencies and draw up an energy balance for the test in MJ/min and in percentages.
(Ans. 31·8 kW, 24·4 kW, 29·6%, 22·7%, 6·454; 1·465, 22·7%; 0·88, 13·62%; 4·106, 63·68%)

9. The following results were obtained during a trial on a single-cylinder gas engine: cylinder bore 250 mm, stroke 500 mm, rev/min 235, explosions/min 80·5, m.e.p. 450 kN/m², b.p. 8 kW, gas consumption/min 0·145 m³ at 17°C and pressure of 60 mm water above atmosphere, barometric pressure 750 mm mercury, cooling water flow 20·4 kg/min, temperature rise of cooling water 15 degC, calorific value of gas at s.t.p. = 18 MJ/m³. Calculate the indicated and brake thermal efficiencies, and draw up an energy balance for the test in MJ/min and in percentages. (Take specific gravity of mercury 13·6.)
(Ans. 36·28%, 19·57%, 2·462 MJ; 0·48 MJ, 19·5%; 1·28, 52%; 0·7 MJ, 28·5%)

10. Sketch a typical indicator diagram for a four-stroke cycle petrol engine, and explain briefly each phase of the cycle. The brake m.e.p. of such an engine is 0·8 MN/m², and the specific consumption is 0·34 kg/kWh (brake basis), the calorific value of the fuel being 43 MJ/kg. The capacity of the engine is 880 cc. Find (a) the b.p. developed at a speed of 2500 rev/min; (b) the overall thermal efficiency.

(Ans. 14·67 kW, 24·62%)

11. The intake of gas and air to a four-stroke engine is 0·85 of the swept volume and the c.v. of the mixture of gas and air is 2 MJ/m³ under the conditions of intake. The engine is to develop i.p. of 30 kW when running at 180 rev/min and firing on 90% of the cycles. The estimated indicated thermal efficiency is 30%, and the ratio $\frac{\text{diameter}}{\text{stroke}} = \frac{1}{1\cdot75}$. Determine the necessary cylinder diameter and stroke and the probable indicated m.e.p. in kN/m².

(Ans. 305·6 mm, 534·7 mm, 567 kN/m²)

12. A small experimental single-cylinder oil engine has a bore of 120 mm and a stroke of 140 mm, and works on the four-stroke cycle; the speed is 600 rev/min and the mean effective pressure is 550 kN/m². The engine uses 1·2 kg of oil per hour of calorific value 45 MJ/kg. The cylinder-jacket cooling water enters at a temperature of 17°C and leaves at 60°C, the quantity being 109 kg/h. The brake wheel diameter is 0·85 m, the net load is 11·2 kg, and the diameter of the rope is 20 mm.

(a) Find the thermal and mechanical efficiencies.

(b) Draw up an energy account for the engine in MJ/min assuming that 6% of the heat supplied in the fuel is lost to the environment.

(Ans. 20%, 29%, 69%, 0·9, 0·18, 0·327, 0·339 MJ)

13. A four-cylinder petrol engine running at 2800 rev/min gave brake power of 20 kW, and when one cylinder was cut out the average torque was 48·1 Nm at the same speed. For normal running at this speed the engine used 0·117 kg/min of petrol of calorific value 44 MJ/kg. Find the mechanical efficiency, the specific fuel consumption brake basis and the indicated thermal efficiency.

(Ans. 84·74%, 0·351 kg/kWh, 27·5%)

14. The following observations were recorded during a trial of a four-stroke single-cylinder oil engine.

Duration of trial	30 min
Oil consumption	4·4 kg
Calorific value of oil	42 MJ/kg
Average area of indicator diagram	850 mm²
Length of diagram	80 mm
Spring scale	56 kN/m² per mm
Brake wheel diameter	1·5 m
Speed	200 rev/min
Brake load	135 kgf
Spring balance reading	18 kgf

Cylinder diameter	300 mm
Stroke	450 mm
Cooling water	11 kg/min
Temperature rise of cooling water	36 degC

1. Calculate: (a) i.p.; (b) b.p.; (c) mechanical efficiency; (d) specific fuel consumption; (e) indicated thermal efficiency.
2. Draw up a heat balance sheet on a 1-min basis.

(Ans. 31·54, 18·03, 57·2%, 0·488 kg/kWh, 30·73%, 1·082 MJ, 1·658 MJ, 3·420 MJ)

[U.E.I.]

15. The following observations were made during the trial of a gas engine having a cylinder 150 mm diameter and a stroke of 220 mm.

Speed, 300 rev/min; firing cycles, 138/min; mean effective pressure, 480 kN/m²; net brake load, 19 kgf; effective brake wheel diameter, 1·3 m; gas consumption, 70 dm³/min at a pressure of 748 mm of mercury and 20°C; calorific value of the gas, 16·5 MJ/m³ at s.t.p. (0°C and 760 mm of mercury); jacket water, 2·1 kg/min; temperature rise of jacket water, 49°C.

Calculate: (a) the gas consumption at s.t.p. per indicated power output (kWh); (b) the indicated thermal efficiency. Also draw up a heat balance in kJ per minute. The heat carried away by the exhaust was 35% of the heat supplied to the engine.

(Ans. (a) 64·19 dm³/min, (b) 24·32%, (c) work 228·4, cooling 430·8, exhaust, 370·8, others 29·0)

[U.L.C.I.]

16. During the trial of a petrol engine, the following results were obtained: duration of test, 1 hour; i.p. = 36·8 kW; b.p. = 31·0 kW; petrol consumption 9·8 kg of calorific value 42 MJ/kg; cooling water passed through cylinder cooling jackets = 450 kg, raised in temperature from 15°C at inlet to 72°C at outlet. The exhaust gases leaving the cylinder were passed through a heat exchanger in which 800 kg of water were raised from 15°C to 64°C.

Determine the mechanical and brake thermal efficiencies and draw up a heat balance sheet (expressed as MJ per minute) for the engine.

(Ans. 84·26%; 27·12%; supply 6·860; work 1·860; cooling 1·790; heat exchanger 2·735; others 0·475)

[U.L.C.I.]

17. The following data refer to a test carried out at constant load and constant speed on a single-cylinder, four-stroke oil engine:

duration of test = 20 min; mean area of typical indicator card = 780 mm^2; length = 75 mm; indicator spring stiffness = 90 kN/m^2/mm of diagram height; cylinder bore = 200 mm; stroke = 350 mm; speed = 320 rev/min; net brake load = 110 kgf at 0·6 m radius; oil consumption = 1·87 kg of calorific value = 45 MJ/kg; cylinder cooling-water flow = 8·5 kg/min, with inlet temperature = 14°C, outlet temperature = 51°C.

Calculate the indicated and brake powers, the mechanical efficiency and the brake thermal efficiency of the engine. Draw up the heat balance for this test, expressed as MJ per minute and as percentages of the heat supplied to the engine.

(Ans. 27·45 kW, 21·7 kW; 79·07%, 30·94%; supply 4·207 (100%); work 1·302 (30·9%); cooling 1·317 (31·3%); un-measured 1·588 (35·8%)
[U.L.C.I.]

18. During a test carried out on a four-cylinder, four-stroke petrol engine, the following readings were recorded: speed 2800 rev/min; indicated mean effective pressure 860 kN/m^2; net brake load 20·2 kgf at 0·5 m radius; fuel consumption 0·17 kg per min, of calorific value 43 MJ/kg; cooling-water flow 10 kg per min, with a temperature rise of 53°C; the heat carried away in the exhaust gas was estimated to be 2·2 MJ per min. The cylinder dimensions were 76 mm bore and 100 mm stroke.

Determine the brake and indicated powers, the fuel consumption in kg/kWh (brake basis), and draw up a heat balance for the test, each item being expressed in MJ per minute, and as a percentage of the heat supplied by the combustion of the fuel.

(Ans. 29·06 kW, 36·42 kW; 0·351 kg/kWh (brake); supply 7·31 Btu = 100%; work 1·744 = 23·9%, cooling 2·219 = 30·2%, exhaust 2·200 = 30·1%, unmeasured 1·147 = 15·8%)
[U.L.C.I.]

19. During a test of a single-cylinder, four-stroke oil engine the following results were recorded: cylinder bore 215 mm, stroke 356 mm; indicator diagram: area 320 mm^2, length 75 mm, spring stiffness 140 kN/m^2 per mm of diagram height; speed 300 rev/min; brake-load 45 kg at 1 m radius; fuel consumption 3·8 kg/h, of calorific value 43 MJ/kg; cooling water flow 3·7 kg/min raised in temperature from 13°C to 60°C.

Determine (a) the mechanical efficiency, and (b) the indicated thermal efficiency of the engine. Draw up a heat balance sheet expressed as MJ/min and as percentages of the heat supplied to the engine. Explain why the friction power has been included in or omitted from your heat balance.

(Ans. 71·9%, 42·5%; supply = 2·724 = 100%, work 0·8322 = 30·6%, cooling 0·7281 = 26·7%, unmeasured 1·1637 = 42·7%)

[U.E.I.]

20. A single-cylinder, four-stroke cycle, gas engine, governed by the hit-and-miss method, is tested at constant load and constant speed over a period of 20 minutes. The following readings were recorded: bore 180 mm; stroke 250 mm; speed 360 rev/min; firing-cycles/min = 145; area of indicator card 770 mm², length 95 mm, when using a spring of stiffness 100 kN/m² per mm of diagram height; net brake load 34 kgf at an effective radius of 0·8 m; gas consumption = 2·33 standard cubic metre, of calorific value = 18 MJ/standard cubic metre; cooling-water flow = 67 kg raised in temperature from 9°C to 50°C.

Determine the indicated and brake powers, and the mechanical efficiency of the engine. Draw up a heat balance in MJ/min, expressing each item as a percentage of the heat input to the engine from the gas.

(Ans. 12·48 kW, 10·09 kW; 80·7%, supply 2·097 = 100%, work 0·603 = 28·7%, cooling 0·575 = 27·3%, unmeasured 0·919 = 44%)

[U.E.I.]

21. Sketch and describe, in simple detail, the essential features of a petrol-engine carburettor featuring either constant depression or a constant choke. Show how the carburettor supplies a richer mixture for cold-starting purposes.

[U.E.I.]

22. During a test of a single-cylinder, four-stroke cycle gas engine, governed by the hit-and-miss method, the following readings were obtained: speed 320 rev/min; firing strokes/min 110; indicator card area, 1000 mm²; length 75 mm, when using a spring stiffness 55 kN/m² of diagram height; net brake load, 30 kgf at 0·75 m radius; gas consumption, 4·95 m³/h when measured at meter conditions of 75 mm of water-gauge and 16°C; c.v. of gas, 18 MJ/m³ at s.t.p.; barometer reading, 745 mm of mercury; cylinder bore, 180 mm; stroke, 265 mm.

Determine the i.p., b.p., mechanical efficiency, and gas consumption in m³/kWh (indicated basis). What is the indicated thermal efficiency of the engine?

1 m³ is measured at conditions of 760 mm of mercury and 0°C. Density of mercury = 13·6 × density of water.

(Ans. 9·067 kW; 7·398 kW; 81·6%; 0·5093 m³/kWh (Ind); 39·26%)

[U.L.C.I.]

23. Sketch a typical indicator diagram which would be obtained from a gas engine working on the four-stroke cycle. Compare this diagram with the theoretical indicator diagram for such an engine.

A six-cylinder, four-stroke petrol engine, of compression ratio 9 : 1 develops 82 kW at 5200 rev/min. The cylinder dimensions are bore 76 mm, stroke 76 mm. Determine the brake mean effective pressure of the engine. If petrol of calorific value 43 MJ/kg is being used at the rate of 0·33 kg/kWh (brake basis), determine (a) the brake thermal efficiency, and (b) the efficiency ratio (or relative efficiency) for the engine.

Take $\gamma = 1\cdot4$.

(Ans. 914·7 kN/m²; 25·37%; 0·434)

[U.L.C.I.]

24. During a constant-load, constant-speed test carried out on a single-cylinder, four-stroke oil engine, the following readings were recorded: speed 360 rev/min; indicated m.e.p., 980 kN/m²; net brake load, 115 kgf at 600 mm radius; oil consumption, 0·23 kg/kWh (Ind); calorific value of oil, 44 MJ/kg; cooling-water flow through engine, 9·0 kg/min with a temperature rise of 39°C; cylinder bore, 215 mm; stroke, 300 mm.

Calculate the i.p. and b.p., mechanical efficiency, and brake thermal efficiency of the engine. Draw up a heat balance for this test, expressing each item in MJ/min and as a percentage of the heat supplied by the fuel.

(Ans. 32·02 kW; 25·53 kW; 79·73%; 28·4%; supplied 5·4 MJ = 100%, work 1·532 MJ = 28·4%, cooling 1·47 MJ = 27·2%, unmeasured 2·398 MJ = 44·4%)

[U.L.C.I.]

4. The Expansion and Compression of Gases

We have seen that work is obtained from an internal combustion engine by allowing a gas, which has been heated to a high pressure, to expand against a piston and drive it forward. It is, therefore, necessary now to examine the behaviour of a gas when it expands in a cylinder, so as to be able to obtain the best advantage from it, and to understand the ensuing practical problems.

When a gas is expanded or compressed in an engine cylinder, it follows the law $pV^n = c$ where c is a constant. It is fairly easy to check this by taking any expansion or compression line from an indicator diagram, and plotting from it a graph of corresponding values of $\log p$ against $\log V$.

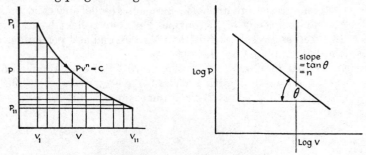

Fig. 35. Expansion and Compression Diagram

If the graph is a straight line, then the expansion or compression curve must be of the form $pV^n = c$, since

$$\text{If } pV^n = c$$

Taking logs $\quad \log p + n \log V = \log c$

$$\therefore \log p = -n \log V + \log c$$

which is of the form $\quad y = mx + c$

i.e. A straight line of slope $-n$.

72

Fig. 36 shows a variety of ways in which a gas may expand according to the law $pV^n = c$ from an initial pressure p_1 in the clearance volume, through the stroke to the final volume V_2.

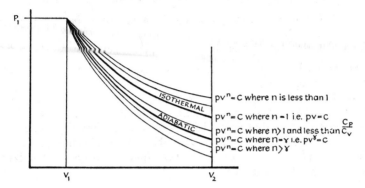

$$pv^n = c \text{ where } n \text{ is less than } 1$$
$$pv^n = c \text{ where } n = 1 \text{ i.e. } pv = c$$
$$pv^n = c \text{ where } n > 1 \text{ and less than } \frac{C_p}{C_v}$$
$$pv^n = c \text{ where } n = \gamma \text{ i.e. } pv^\gamma = c$$
$$pv^n = c \text{ where } n > \gamma$$

Fig. 36. Isothermal and Adiabatic Expansion and Compression
of a Gas

There are an infinite number of such expansions, each with a different value of n, but two of them are of particular importance, the ISOTHERMAL expansion and the ADIABATIC expansion.

THE ISOTHERMAL EXPANSION OR COMPRESSION OF A GAS (pV=c)

That particular expansion having a value of $n = 1$ is called an ISOTHERMAL expansion.

If $n = 1$, then $pV = c$.

But for a gas, the characteristic equation states $pV = mRT$

$$\therefore \quad mRT = c \text{ when } n = 1.$$

Now since the mass of the gas remains constant, and R is the characteristic gas constant, then during this process T must be constant.

∴ An ISOTHERMAL expansion or compression of a gas is one in which the temperature of the gas remains constant throughout the process.
In practice, an isothermal expansion is approached when the operation is carried out very slowly so that heat can be taken in and the temperature remain constant.

73

THE ADIABATIC EXPANSION OR COMPRESSION OF A GAS ($pV^{\gamma}=c$)

There will be, for any particular gas, an expansion having a value of n which exactly equals the ratio of the specific heats $\dfrac{c_p}{c_v}$ ($=\gamma$). That particular expansion, having a value of $n=\gamma$, is called an ADIABATIC expansion.

By definition an adiabatic expansion is one in which no heat flow takes place across the boundaries of the system. Thus it will be necessary to prove that no heat flow takes place when $n=\gamma$, and this is done on page 79. In practice, an adiabatic expansion is approached when the operation is carried out very quickly so that there is no time for heat flow to occur.

A POLYTROPIC EXPANSION OR COMPRESSION OF A GAS ($pV^{n}=c$)

All expansions or compressions of the form $pV^n=c$ other than the particular isothermal and adiabatic operations are called 'polytropic' operations. Fig. 36 shows a sample of polytropic expansions, those lying above the isothermal curve having a value of n less than 1, those lying between the isothermal and adiabatic curves having a value of n greater than 1 but less than γ, and those lying below the adiabatic curve having a value of n greater than γ.

Since for a polytropic process $pV^n=c$, and also for a gas the characteristic equation gives $\dfrac{pV}{T}=$ constant, we can obtain a useful relationship between temperatures and pressures, or temperatures and volumes, for any two points during the process as follows:—

$$pV^n=c \quad \therefore \quad p_1V_1{}^n=p_2V_2{}^n \qquad \ldots[1]$$

and also

$$\frac{pV}{T}=c \quad \therefore \quad \frac{p_1V_1}{T_1}=\frac{p_2V_2}{T_2} \qquad \ldots[2]$$

$$\text{From [2]} \quad \frac{T_2}{T_1}=\frac{p_2V_2}{p_1V_1} \qquad \ldots[3]$$

To eliminate volumes, from equation [1] $\left(\dfrac{V_2}{V_1}\right)^n=\dfrac{p_1}{p_2}$

$$\therefore \frac{V_2}{V_1} = \left(\frac{p_1}{p_2}\right)^{\frac{1}{n}}$$

$$= \left(\frac{p_2}{p_1}\right)^{-\frac{1}{n}}$$

and substituting in [3]

$$\frac{T_2}{T_1} = \frac{p_2}{p_1} \times \left(\frac{p_2}{p_1}\right)^{-\frac{1}{n}}$$

$$\therefore \frac{T_2}{T_1} = \left(\frac{p_2}{p_1}\right)^{\frac{n-1}{n}}$$

Or eliminating pressures, from equation [1], $\frac{p_2}{p_1} = \left(\frac{V_1}{V_2}\right)^n$ and substituting in [3]

$$\frac{T_2}{T_1} = \left(\frac{V_1}{V_2}\right)^n \times \frac{V_2}{V_1}$$

$$= \left(\frac{V_1}{V_2}\right)^n \times \left(\frac{V_1}{V_2}\right)^{-1}$$

$$\therefore \frac{T_2}{T_1} = \left(\frac{V_1}{V_2}\right)^{n-1}$$

Giving the combined relationship for any polytropic process

$$\frac{T_2}{T_1} = \left(\frac{p_2}{p_1}\right)^{\frac{n-1}{n}} = \left(\frac{V_1}{V_2}\right)^{n-1}$$

This very useful equation is well worth committing to memory.

WORK DONE DURING AN EXPANSION OR COMPRESSION—NON-FLOW PROCESSES

Consider a non-flow process in a closed system in which a gas initially at pressure p_1 and volume V_1 expands to a volume V_2 and corresponding pressure p_2. The work done during the expansion is given by the area under the expansion curve on a pV diagram. Since the area is an irregular one, we must use some summation method for measuring it, and the most convenient is to use the calculus, which gives

$$\text{Work done} = \int_{V_1}^{V_2} p.dV$$

75

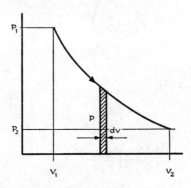

Fig. 37. Work Done during Expansion

i.e. the sum of all the strips of area $p.dV$ lying between V_1 and V_2. We cannot progress further than this, since p is varying, without knowing the law of the expansion.

(a) Work done during a polytropic expansion

For a polytropic process $pV^n = c$

$$\therefore p = \frac{c}{V^n}$$

$$\therefore \text{Work done} = \int_{V_1}^{V_2} p.dV.$$

$$= c \int_{V_1}^{V_2} \frac{dV}{V^n}$$

$$= c \left[\frac{V^{1-n}}{1-n} \right]_{V_1}^{V_2}$$

$$= \frac{c}{1-n} [V_2^{1-n} - V_1^{1-n}]$$

Now $c = p_1 V_1^n = p_2 V_2^n$, and $p_2 V_2^n \times V_2^{1-n} = p_2 V_2$;
and $p_1 V_1^n \times V_1^{1-n} = p_1 V_1$;

$$\therefore W = \frac{1}{1-n} (p_2 V_2 - p_1 V_1)$$

$$\therefore W = \frac{p_1 V_1 - p_2 V_2}{n-1}$$

76

(b) Work done during an isothermal expansion

For an isothermal process, $pV = c$, i.e. $n = 1$.

Note that if we try to substitute in the general work equation for a $pV^n = c$ curve, since $p_1 V_1 = p_2 V_2$, for an isothermal process, the equation becomes $W = \dfrac{p_1 V_1 - p_2 V_2}{n - 1}$

$$\therefore \ W = \frac{0}{0} \ \text{which we cannot interpret.}$$

Hence it is necessary to start from the fundamental work equation

$$W = \int_{V_1}^{V_2} p\, dV$$

For an isothermal process $pV = c$

$$\therefore \ p = \frac{c}{V}$$

$$\therefore \ \text{Work done} = \int_{V_1}^{V_2} p\, dV$$

$$= c \int_{V}^{V_2} \frac{dV}{V}$$

$$= c \left[\log_e V \right]_{V_1}^{V_2}$$

and $c = p_1 V_1 = p_2 V_2$

$$\therefore \ W = p_1 V_1 \log_e \frac{V_2}{V_1}$$

(c) Work done during an adiabatic expansion

For an adiabatic process $pV^\gamma = c$

Now for a polytropic process $pV^n = c$, $W = \dfrac{p_1 V_1 - p_2 V_2}{n - 1}$

$\therefore$ for an adiabatic process $pV^\gamma = c$, $W = \dfrac{p_1 V_1 - p_2 V_2}{\gamma - 1}$

To prove that no heat flow occurs across the boundaries of a system (i.e. the process is adiabatic) during an expansion of the form $pV^n = c$ when $n = \gamma$.

Let a gas expand according to $pV^n = c$ from p_1 and V_1 to p_2 and V_2 without taking in or rejecting heat.

From the First Law for a closed system $Q = W + (U_2 - U_1)$, and for no heat flow $Q = 0$

$$\therefore \ 0 = \frac{p_1 V_1 - p_2 V_2}{(n - 1)} + mc_v(T_2 - T_1)$$

Now $p_1 V_1 = mRT_1$ and $p_2 V_2 = mRT_2$.

$$\therefore \quad 0 = mR \cdot \frac{(T_1 - T_2)}{n-1} + mc_v (T_2 - T_1)$$

$$\therefore \quad mR (T_2 - T_1) = (n-1) \, mc_v (T_2 - T_1)$$

Now $R = c_p - c_v$

$$\therefore \quad c_p - c_v = nc_v - c_v$$

$$\therefore \quad n = \frac{c_p}{c_v} = \gamma$$

Hence during an expansion of the form $pV = c$, no heat flow takes place, and the expansion is termed adiabatic.

HEAT FLOW DURING EXPANSION OR COMPRESSION OF A GAS IN A CLOSED SYSTEM

For a closed system being conducted through a non-flow process the First Law states:—

Heat taken in   Work done by the gas   Increase in internal energy

$$Q \qquad = \qquad W \qquad + \qquad (U_2 - U_1)$$

For a polytropic process of the form $pV^n = c$:—

$$Q = W + (U_2 - U_1)$$

$$= \frac{p_1 V_1 - p_2 V_2}{(n-1)} + mc_v (T_2 - T_1)$$

and since $mT_2 = \dfrac{p_2 V_2}{R}$ and $mT_1 = \dfrac{p_1 V_1}{R}$

$$Q = \frac{p_1 V_1 - p_2 V_2}{(n-1)} + \frac{c_v}{R}(p_2 V_2 - p_1 V_1)$$

and $c_p - c_v = R$

$$\therefore \quad c_v (\gamma - 1) = R \qquad \qquad \therefore \quad \frac{c_v}{R} = \frac{1}{(\gamma - 1)}$$

Hence $\quad Q = \dfrac{p_1 V_1 - p_2 V_2}{(n-1)} - \dfrac{1}{(\gamma - 1)} \cdot (p_1 V_1 - p_2 V_2)$

$$= \frac{p_1 V_1 - p_2 V_2}{(n-1)} \left[1 - \frac{n-1}{\gamma - 1} \right]$$

$$= \frac{\gamma - n}{\gamma - 1} \times \frac{p_1 V_1 - p_2 V_2}{(n-1)}$$

$$\text{i.e. } Q = \frac{\gamma - n}{\gamma - 1} \times \frac{p_1 V_1 - p_2 V_2}{(n - 1)}$$

$$Q = \frac{\gamma - n}{\gamma - 1} \times \textbf{work done}$$

We can summarize the results so far obtained in tabular form as follows:—

	Heating at constant volume	Heating at constant pressure	Polytropic process $pV^n = c$	Isothermal process $pV = c$	Adiabatic process $pV^\gamma = c$
Change of internal energy $U_2 - U_1$	$mc_v(T_2 - T_1)$	$mc_v(T_2 - T_1)$	$mc_v(T_2 - T_1)$	Since $T_2 = T_1$ $(U_2 - U_1) = 0$	$mc\,(T_2 - T_1)$
Work done W	Since no movement $W = 0$	$p(V_2 - V_1)$	$\dfrac{p_1 V_1 - p_2 V_2}{(n-1)}$	$p_1 V_1 \log_e \dfrac{V_2}{V_1}$	$\dfrac{p_1 V_1 - p_2 V_2}{(\gamma - 1)}$
Heat exchange $Q = W + (U_2 - U_1)$	$mc_v(T_2 - T_1)$	$p(V_2 - V_1) + mc_v(T_2 - T_1)$	$\dfrac{\gamma - n}{\gamma - 1} \times$ Work done	$p_1 V_1 \log_e \dfrac{V_2}{V_1}$	0

Example 14

'A petrol engine has a cylinder diameter of 95 mm and stroke 127 mm, clearance volume 230 cm^3. If the temperature at the beginning of compression is 57°C, find the temperature at the end of compression, and the work done during the compression stroke if the law of compression is $pV^{1\cdot 3} = c$. Take the initial pressure as 100 kN/m^2.'

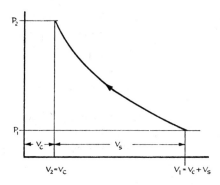

Fig. 38.

$$V_2 = V_c = 230 \times 10^{-6} \text{ m}^3$$
$$V_1 = V_s + V_c$$
$$V_1 = \left(\frac{\pi}{4} \times 0\cdot095^2 \times 0\cdot127\right) + 230 \times 10^{-6}$$
$$= 1130 \times 10^{-6} \text{ m}^3$$

79

$$p_2 V_2{}^{1 \cdot 3} = p_1 V_1{}^{1 \cdot 3} \qquad\qquad \therefore \; p_2 = 100 \left(\frac{V_1}{V_2}\right)^{1 \cdot 3}$$

$$= 10^2 \left(\frac{1130}{230}\right)^{1 \cdot 3}$$

$$= 792 \cdot 1 \;\text{kN/m}^2$$

Now $\dfrac{p_1 V_1}{T_1} = \dfrac{p_2 V_2}{T_2}$

$\therefore$ temperature at end of compression $T_2 = \dfrac{p_2 V_2}{p_1 V_1} \times T_1$

$$= \frac{792 \cdot 1 \times 230}{100 \times 1130} \times 330$$

$$= 532 \;\text{K or } 259°\text{C}$$

Work done during compression stroke

$$W = \frac{p_1 V_1 - p_2 V_2}{n-1} = \frac{100 \times 10^3 \times 1130 \times 10^{-6} - 792 \cdot 1 \times 10^3 \times 230 \times 10^{-6}}{[1 \cdot 3 - 1]}$$

$$= -230 \cdot 6 \;\text{J}$$

(Negative sign indicates work done on the gas.)

Example 15

'1 m³ of air, initially at 110 kN/m² and 15°C, is compressed according to the law $pV^{1 \cdot 3} = $ constant in a cylinder to a final pressure of 1·4 MN/m². Taking R for air $= 287$ J/kgK and $c_p = 1005$ J/kgK, determine

(a) the volume and temperature of the air at the end of the compression;

(b) the work done in compressing the air;

(c) the change of internal energy;

(d) the heat exchange through the cylinder walls, stating the direction of heat flow.'   [U.E.I.]

(a) $p_1 V_1{}^{1 \cdot 3} = p_2 V_2{}^{1 \cdot 3}$

$$\therefore \; V_2 = V_1 \left(\frac{p_1}{p_2}\right)^{\frac{1}{1 \cdot 3}} = 1 \left(\frac{110}{1400}\right)^{0 \cdot 77} = \frac{1}{(12 \cdot 73)^{0 \cdot 77}}$$

$$\doteqdot 0 \cdot 141 \;\text{m}^3$$

$$T_2 = \frac{p_2 V_2}{p_1 V_1} T_1 = \frac{1400 \times 0 \cdot 141}{110 \times 1} \times 288 = 517 \;\text{K or } 244°\text{C}$$

(b) Work done $= \dfrac{p_1 V_2 - p_2 V_2}{n-1} = \dfrac{110 \times 10^3 \times 1 - 1400 \times 10^3 \times 0\cdot141}{0\cdot3}$

$$= -291\cdot3 \text{ kJ}$$

(c) $(U_2 - U_1) = m c_v [T_2 - T_1]$

$$m = \frac{pV}{RT} = \frac{110 \times 10^3 \times 1}{287 \times 288} = 1\cdot33 \text{ kg}$$

$$c_p - c_v = R \qquad \therefore c_v = 1005 - 287$$
$$= 718 \text{ J/kgK}$$

$\therefore$ Change of internal energy $(U_2 - U_1) = 1\cdot33 \times 718 \ (517 - 288)$
$$= 218\cdot8 \text{ kJ}$$

(d) Heat flow $Q = W + (U_2 - U_1)$
$$= -291\cdot3 + 218\cdot8$$
$$= -72\cdot5 \text{ kJ}$$

Negative sign indicates that heat is rejected, i.e. direction of heat flow is outwards from the cylinder.

Example 16

'One kilogramme of a certain gas expands adiabatically in a closed system until its pressure is halved. During the expansion the gas does 67 kJ of external work and its temperature falls from 240°C to 145°C. Calculate the value of the adiabatic index and the characteristic constant of the gas.'

$$\frac{T_2}{T_1} = \left(\frac{p_2}{p_1}\right)^{\frac{\gamma-1}{\gamma}}$$

$\therefore$ Taking logs and rearranging, $\dfrac{\gamma-1}{\gamma} = \dfrac{\log\dfrac{T_1}{T_2}}{\log\dfrac{p_1}{p}}$

$$= \frac{\log\dfrac{513}{418}}{\log 2}$$

$$= 0\cdot2954$$

$$\therefore \ \gamma = 1\cdot421$$

For an adiabatic expansion, heat flow is zero

$$\text{i.e. } 0 = W + (U_2 - U_1)$$
$$\therefore \ 0 = W + mc_v(T_2 - T_1)$$
$$\therefore \ 0 = 67 \times 10^3 + c_v(418 - 513)$$
$$\therefore \ c_v = 705\cdot3 \ \text{J/kgK}$$

From $c_p - c_v = R$

Dividing by c_v, $(\gamma - 1) = \dfrac{R}{c_v}$

$$\therefore \ R = c_v(\gamma - 1)$$
$$= 705\cdot3(1\cdot421 - 1)$$
$$= 296\cdot7 \ \text{J/kgK}$$

Example 17

'In a closed system, one kilogramme of air initially at 100 kN/m² and 27°C is compressed adiabatically to 3 MN/m², and then expanded isothermally back to its original volume. Determine the excess of work done by the gas over work done on the gas. Take $R = 287$ J/kgK and $\gamma = 1\cdot4$ for air.'

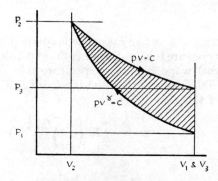

Fig. 39. Excess of Work Done

$$T_2 = T_1\left(\frac{p_2}{p_1}\right)^{\frac{n-1}{n}}$$

$$= 300\left(\frac{3000}{100}\right)^{\frac{0\cdot4}{1\cdot4}}$$

$$= 792\cdot5 \ \text{K}$$

82

Work done during compression $= \dfrac{p_1 V_1 - p_2 V_2}{\gamma - 1}$

$$= \dfrac{mR\,(T_1 - T_2)}{\gamma - 1}$$

$$= \dfrac{1 \times 287(792 \cdot 5 - 300)}{0 \cdot 4} = 353 \cdot 4 \text{ kJ}$$

Work done during isothermal expansion $= p_2 V_2 \, \log_e \dfrac{V_3}{V_2}$

From compression $1 - 2$,

$$p_1 V_1{}^\gamma = p_2 V_2{}^\gamma$$

$$\therefore \ \dfrac{V_1}{V_2} = \left(\dfrac{p_2}{p_1}\right)^{\frac{1}{\gamma}} = \left(\dfrac{3000}{100}\right)^{\frac{1}{1 \cdot 4}} = 11 \cdot 35$$

and $\dfrac{V_1}{V_2} = \dfrac{V_3}{V_2}$ (since $V_1 = V_3$)

$\therefore$ Work done during expansion $= mRT_2 \, \log_e \dfrac{V_3}{V_2}$

$$= 1 \times 287 \times 792 \cdot 5 \log_e 11 \cdot 35$$
$$= 552 \cdot 5 \text{ kJ}$$

$\therefore$ Excess of work done by the gas

$$= 552 \cdot 5 - 353 \cdot 4$$
$$= 199 \cdot 1 \text{ kJ}$$

Example 18

'A volume of 0·5 m³ of gas is expanded in a cylinder from a pressure of 660 kN/m² and temperature 165°C to a pressure of 120 kN/m² according to the law $pV^{1 \cdot 3} = c$. Find:—

(a) the final volume and temperature of the air;

(b) the work done by the air during the expansion;

(c) the change of internal energy during the expansion;

(d) the heat flow across the cylinder walls during the expansion, stating its direction.

Take $c_v = 710$ J/kgK and $R = 287$ J/kgK.

(a) $V_2 = V_1 \left(\dfrac{p_1}{p_2}\right)^{\frac{1}{n}} = 0 \cdot 5 \left(\dfrac{660}{120}\right)^{\frac{1}{1 \cdot 3}} = 1 \cdot 856 \text{ m}^3$

$T_2 = \dfrac{p_2 V_2}{p_1 V_1} \times T_1 = \dfrac{120 \times 1 \cdot 856}{660 \times 0 \cdot 5} \times 438 = \underline{295 \cdot 7 \text{ K}}$

(b) Work done $= \dfrac{p_1 V_1 - p_2 V_2}{n - 1}$

$$= \dfrac{660 \times 10^3 \times 0 \cdot 5 - 120 \times 10^3 \times 1 \cdot 856}{0 \cdot 3} = \underline{357 \cdot 7 \text{ kJ}}$$

(c) $(U_2 - U_1) = mc_v(T_2 - T_1)$

$$m = \frac{pV}{RT} = \frac{660 \times 10^3 \times 0.5}{287 \times 438} = 2.625 \text{ kg}$$

$$\therefore \ (U_2 - U_1) = 2.625 \times 710(295.7 - 438)$$
$$= -265.3 \text{ kJ}$$

(d) $Q = W + (U_2 - U_1)$
$$= 357.7 - 265.3$$
$$= 92.4 \text{ kJ}$$

Positive sign indicates heat taken in.

FLOW PROCESSES

The previous work in this chapter has dealt with non-flow processes, i.e. processes in which the same mass of working agent has remained in the cylinder. In a flow process, however, such as occurs in engines having a steady flow of working agent through them (see definition page 9), the fluid is being propelled through the pipeline against the downstream resistance and FLOW WORK is being done.

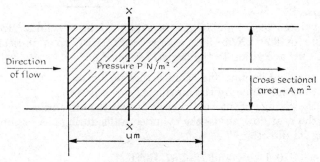

Fig. 40. Flow Work

Flow Work

Consider the flow of a fluid past the section XX of a pipe of cross section A. If the flow velocity is v m/s, and the pressure at the section is p,

Flow work at the section XX per second $= pAv$ J
$$\therefore \text{ Flow work} = pV \text{ W}$$

where $V =$ volume flowing per second.

84

This work is part of the total energy passing the section at any instant, and exists only when the fluid is flowing and is being transferred.

THE GENERAL ENERGY EQUATION

Fig. 41 represents an open system in which a steady-flow process is taking place. At entry to the system the working agent possesses potential, kinetic and internal energy and entry flow work is done. During its passage through the system the working agent is considered to take in a quantity of heat Q and do external work to the value of W. At exit from the system the working agent will again possess potential, kinetic and internal energy and will do flow work to leave the system.

Stored energy at input Section $X_1 = PE_1 + KE_1 + U_1$

In chapter 1, page 19, we have seen that the First Law states for an open system:—

Stored energy at entry + flow-work at entry + heat inflow to system = stored energy at exit + flow-work at exit + work outflow from system

i.e.

Initial potential energy $9.81mZ_1$ Final potential energy $9.81mZ_2$

$+$ Initial kinetic energy $\dfrac{mv_1^2}{2}$ $+$ Final kinetic energy $\dfrac{mv_2^2}{2}$

$+$ Initial internal energy U_1 $=$ $+$ Final internal energy U_2

$+$ Initial flow work p_1V_1 $+$ Final flow work p_2V_2

$+$ Heat input Q $+$ External work done W

$$\therefore 9.81\, mZ_1 + \frac{mv_1^2}{2} + U_1 + p_1V_1 + Q = 9.81\, mZ_2 + \frac{mv_2^2}{2} + U_2 + p_2V_2 + W$$

This is the General Energy Equation for a flow process, which again is such a useful and important statement that it ought to be committed to memory.

Notice that if the process is non-flow, there is no change of potential energy, no change of velocity, and no flow work, and hence the general equation reduces to:—

$$U_1 + Q = U_2 + W$$
$$Q = W + (U_2 - U_1)$$

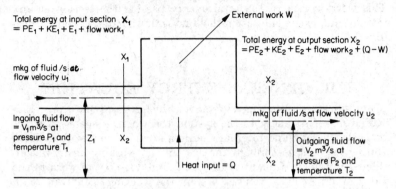

Fig. 41. General Energy Equation

i.e. Q = external work done + change of internal energy which we gave called the 'Energy equation for a closed system'.

ENTHALPY

A very useful property of fluids is known as Enthalpy and is defined by

$$\text{Enthalpy } H = U + pV$$

Hence in the General Energy Equation for a flow process

$$9 \cdot 81 \ mZ_1 + \frac{mv_1^2}{2} + U_1 + p_1 V_1 + Q = 9 \cdot 81 \ mZ_2 + \frac{mv_2^2}{2} + U_2 + p_2 V_2 + W$$

we may write

$$9 \cdot 81 \ mZ_1 + \frac{mv_1^2}{2} + H_1 + Q = 9 \cdot 81 \ mZ_2 + \frac{mv_2^2}{2} + H_2 + W$$

Now, in many flow processes with which we are concerned, the change of potential energy is negligible because of the small changes of level and the small weights involved. Similarly, because the entering and leaving velocities are much the same in many heat engines (e.g. steam turbines) the kinetic energy terms can also be neglected. In such cases, the General Equation reduces to

$$H_1 + Q = H_2 + W$$
$$\text{or } Q = W + (H_2 - H_1)$$

i.e. for flow processes, heat taken in or rejected
= work done + change of enthalpy.

Example 19

'During a steady-flow process in an open system the velocity of

86

the fluid is increased from 100 to 600 m/s, the volume is decreased from 5·0 to 2·0 m³/kg, the pressure is increased from 1 MN/m² to 2 MN/m² abs. and the internal energy is increased by 20 kJ/kg. Find the change of enthalpy per kg of the fluid, and the work done on the fluid if no heat is taken in or rejected. There is no change of potential energy.'

Initial flow work $p_1 V_1 = 1 \times 10^6 \times 5$
$$= 5 \text{ MJ/kg}$$
Final flow work $p_2 V_2 = 2 \times 10^6 \times 2$
$$= 4 \text{ MJ/kg}$$

Hence change of enthalpy/kg of fluid

$$
\begin{aligned}
h_2 - h_1 &= (u_2 + p_2 V_2) - (u_1 + p_1 V_1) \\
&= (u_2 - u_1) + (p_2 V_2 - p_1 V_1) \\
&= 20 \times 10^3 + (4 \times 10^6 - 5 \times 10^6) \\
&= \underline{-980 \text{ kJ}} \text{ (decrease of enthalpy)}
\end{aligned}
$$

From the general energy equation, neglecting the potential energy term; for 1 kg of fluid:—

$$\frac{v_1^2}{2} + u_1 + p_1 V_1 + Q = \frac{v_2^2}{2} + u_2 + p_2 V_2 + W$$

$$-W = (h_2 - h_1) + \tfrac{1}{2}(v_2^2 - v_1^2) \text{ when } Q = 0$$

$$-W = -980 \times 10^3 + \left(\frac{600^2 - 100^2}{2}\right) = -980 \times 10^3 + 175 \times 10^3 \text{ J/kg}$$

$$= \underline{805 \text{ kJ/kg}}$$

Enthalpy of an Ideal Gas

By definition, enthalpy of a fluid is given by

$$H = U + pV$$

Now the internal energy of a gas is a function of temperature only and may be written $U = mc_v T$

Hence, enthalpy $H = U + pV$

$$\therefore H = mc_v T + mRT$$
$$\therefore H = mT(c_v + R)$$
$$\therefore \quad = mc_p T$$

Example 20
'A rotary air compressor deals with 450 kg of natural gas per

87

minute. The gas is taken in at 15°C and 100 kN/m² and is compressed in accordance with the law $pV^{1\cdot25} = $ constant, to 175 kN/m². The power used in compressing the gas is 800 kW. Neglecting the difference in intake and exit gas velocities, how much heat is transferred per minute? Take $c_p = 2\cdot26$ kJ/kgK.'

We may also neglect the change of potential energy, and hence the general equation for this flow process may be written

$$Q = W + (H_2 - H_1)$$

Since power expended $= 800$ kW,

Work done per second $W = -800 \times 10^3 \times 60$ J
$$= -48 \text{ MJ (work done } on \text{ the gas and hence flowing } into \text{ the system)}$$

Change of enthalpy $(H_2 - H_1) = mc_p(T_2 - T_1)$

For adiabatic compression $T_2 = T_1\left(\dfrac{p_2}{p_1}\right)^{\frac{n-1}{n}}$

$$= 288\left(\frac{175}{100}\right)^{\frac{0\cdot25}{1\cdot25}} = 322\cdot1\text{K}$$

Hence change of enthalpy

$$(H_2 - H_1) = mc_p(T_2 - T_1)$$
$$= 450 \times 2260(322\cdot1 - 288)$$
$$= 34\cdot68 \text{ MJ}$$
$$\therefore Q = W + (H_2 - H_1)$$
$$= -48 + 34\cdot68$$
$$= -13\cdot32 \text{ MJ (i.e. heat rejected).}$$

Example 21

'Derive an expression for the work done by a mass of gas in a steady flow process involving frictionless adiabatic expansion where there is no change of kinetic or potential energy.

Oxygen is flowing steadily through an open system at a rate of 10 kg/s from an initial pressure of 9 MN/m² and temperature 200°C, and is expanding adiabatically in the process to a final pressure of 3 MN/m². Assuming no change in kinetic or potential energy and neglecting the effect of friction, calculate:—

(a) the final temperature of the oxygen,
(b) the work flowing from the system per minute.

Take c_p for Oxygen $= 917$ J/kgK; Universal gas constant $= 8314$ J/kmolK.'

Proof:—
From the General Energy Equation, for a flow process with no change of kinetic or potential energy

$$Q = W + (H_2 - H_1) \qquad \ldots [1]$$

For an adiabatic expansion $Q = 0$, and using the expression for enthalpy of a gas $H = mc_pT$

Hence, from (1), $W = mc_p(T_1 - T_2)$

(A case such as this occurs in a stationary gas turbine engine where gas is expanded adiabatically through the turbine with a negligible change in kinetic or potential energy.)

To find γ for Oxygen:—

$$c_p - c_v = R \text{ and } R = \frac{8314}{M}$$

$$\therefore R_{\text{oxygen}} = \frac{8314}{32}$$

$$= 259 \cdot 8 \text{ J/kgK}$$

Hence, $\qquad 917 - c_v = 259 \cdot 8$

$$\therefore c_v = 657 \cdot 2 \text{ J/kgK}$$

$$\therefore \gamma = \frac{c_p}{c_v} = \frac{917 \cdot 0}{657 \cdot 2} = 1 \cdot 395$$

To find final temperature:

$$\frac{T_2}{T_1} = \left(\frac{p_2}{p_1}\right)^{(\gamma-1)/\gamma}$$

$$\therefore T_2 = 473\left(\frac{3}{9}\right)^{0 \cdot 395/1 \cdot 395}$$

$$= 346 \cdot 5 \text{ K}$$

$\therefore$ final temperature $t_2 = 346 \cdot 5 - 273 = \underline{73 \cdot 5 °C}$

To find work flow per minute:

$$W = mc_p(Y_1 - T_2)$$
$$= 10 \times 60 \times 917(473 - 346 \cdot 5) \text{ J}$$
$$= \underline{40 \cdot 44 \text{ MJ}}$$

89

REVISION EXERCISES—CHAPTER 4

1. The compression ratio for a petrol engine of bore 100 mm and stroke 120 mm is 5·5. At the end of the suction stroke the pressure in the cylinder is 95 kN/m², and the temperature 35°C. Calculate the mass of the cylinder contents, taking their molecular weight as 28·8. Compression takes place following the law $pV^{1·15} = c$.

Determine (a) the pressure and temperature of the gas at the end of compression; (b) the work done on the gas during compression.

Take Universal gas constant = 8306 J/kgmolK.

(Ans. 0·0123 kg, 674·8 kN/m², 125°C, – 1·74 kJ)

2. A quantity of gas is expanded in the cylinder of an engine from a pressure of 700 kN/m², volume 2·5 dm³ and temperature 1100°C to a final pressure of 270 kN/m², according to the law $pV^{1·3} = c$. Find

   (a) the final volume and temperature;
   (b) the work done during the expansion;
   (c) the change of internal energy during the expansion.

   Take R for the gas = 289 J/kg degC, c_v = 710 J/kg degC.

   (Ans. 0·0052 dm³, 829°C, 1·153 kJ, – 842·4 J)

3. 0·05 kg of air at a temperature of 37°C and pressure 100 kN/m² is compressed in a closed system to a pressure of 500 kN/m² abs. according to the law $pV^{1·3} = c$. Given that $c_p = 1005$ J/kg degC and $c_v = 708$ J/kg degC for air, calculate

   (a) the work done during compression;
   (b) the change in internal energy of the air;
   (c) the heat flow stating whether this is a rejection or reception.

   (Ans. – 6·667 kJ, 4·956 kJ, 1·711 kJ rejected)

4. 0·03 m³ of a gas, initially at a pressure of 1·4 MN/m², temperature 150°C, is expanded in a closed system to a volume of 0·2 m³ according to the law $pV^{1·2} = c$. Determine (a) the mass of the gas; (b) the change in the internal energy; (c) the heat flow through the cylinder walls stating whether this is a gain or a loss.

   Take $R = 640$ J/kg degC, and $c_p = 2$ kJ/kg degC.

   (Ans. 0·155 kg, – 28·16 kJ; 38·84 kJ gain)

5. State the difference between an adiabatic and an isothermal process. The total volume at the end of the induction stroke of a

petrol engine is 3 dm³, the pressure and temperature then being 100 kN/m² and 82°C respectively. Adiabatic compression follows until the volume is reduced to one-sixth of its original volume, when 2·5 kJ is added by combustion at constant volume. Given that $c_v = 708$ and $R = 287$ J/kg degC for air, calculate

    (a) the mass of charge induced into the cylinder assuming this to have the characteristics of air;

    (b) the temperature at the end of compression;

    (c) the temperature and pressure at the end of combustion.

(Ans. 0·002 94 kg, 480°C, 1 679°C, 3·3 MN/m²)

6. 1 m³ of air at 110 kN/m² and 15°C is compressed adiabatically in a closed system to one-quarter of its original volume. The air is then cooled at constant pressure until the temperature is again 15°C. The air is then expanded back to its original conditions of 110 kN/m², 1 m³ and 15°C. Calculate (a) the temperature and pressure at the end of compression (assume $\gamma = 1·4$); (b) the volume at the end of heat rejection; (c) the work done during compression and constant pressure cooling, and (d) the index of expansion during the expansion back to the original condition.

(Ans. 229°C and 766 kN/m², 0·143 m³, 82 kJ, 203·8 kJ, $n = 1$)

7. In the characteristic equation $pV = RT$ for unit mass of a gas, prove that R is equal to the difference between the specific heats at constant pressure and constant volume. One kilogramme of nitrogen at 15°C is compressed adiabatically in a closed system to one-quarter of its volume and the temperature is then 228°C. The work done on the gas during compression is 157·5 kJ. Calculate the specific heats at constant pressure and constant volume and the gas constant R.

(Ans. 1035; 740; 295)
[L.U.]

8. Prove that the index γ in $pV^{\gamma} = $ constant, for the adiabatic expansion of a gas is the ratio of the specific heat at constant pressure to the specific heat at constant volume. A certain gas occupies 4 m³ at 110 kN/m² and 20°C. It is compressed adiabatically in a closed system to a pressure of 690 kN/m². Find the new temperature, the new volume, and the change in internal energy of the gas, given that the density is 1·39 kg/m² at s.t.p. and the specific heat at constant volume is 732 J/kg degC.

(Ans. 206°C; 1·042 m³; 766 kJ)
[L.U.]

9. Two points chosen on the compression curve of an indicator card

91

from an air compressor are found to scale pressure 350 kN/m², volume 10 dm³, and pressure 110 kN/m², volume 27 dm³ respectively. The mass of air under compression is 0·027 kg. For the compression between the two chosen points find

(a) the index n, if the law of compression is $pV^n = \text{constant}$;
(b) the rise in temperature;
(c) the work done on the air.
Take $R = 287$ J/kg degC.

(Ans. 1·165, 68·6°C, 3·21 kJ)
[L.U.]

10. The pressure and temperature of the air in a cylinder are 95 kN/m² and 38°C. The air is compressed according to the law $pV^{1·24} = \text{constant}$, until the pressure is 650 kN/m². The volume of air initially is 1·25 m³. Find

(a) the mass of air in the cylinder;
(b) the temperature at the end of compression;
(c) the work done on the air during compression;
(d) the heat rejected by the air during compression.
Take $R = 287$ J/kg degC. $\gamma = 1·4$.

(Ans. 1·33 kg, 179°C, 214 kJ, 81·4 kJ)

11. 0·34 m³ of gas at 1 MN/m² and 130°C expands adiabatically until the pressure is 105 kN/m² after which it is compressed isothermally to its original volume. Find the final temperature and pressure of the gas and the change in internal energy. The specific heats are $c_p = 996$ J/kg degC, $c_v = 704$ J/kg degC.

(Ans. −65°C; 517 kN/m² and 396·7 kJ)
[U.E.I.]

12. 1 kg of gas expands adiabatically in a closed system from a pressure of 700 kN/m² and a volume of 0·12 m³ to a final pressure of 150 kN/m². Determine the final volume, given that γ for the gas = 1·38. Find also the work done during the expansion, and taking c_v for the gas as 720 J/kg degC, calculate the change of temperature which occurs.

(Ans. 0·367 m³; 76·4 kJ; 106 degC)

13. 0·9 kg of air are compressed at a constant temperature of 37°C in a cylinder from an initial pressure of 69 kN/m² to a final pressure 565 kN/m². Find the compression ratio and the heat flow during the process.
R for air = 287 J/kg degC.

(Ans. 8·2; −168 kJ)

14. 145 dm³ of air at a pressure of 69 kN/m² and a temperature of 18°C is compressed in a closed system according to the law $pV^{1·18} = c$ to a volume of 50 dm³.

Given γ for air $= 1\cdot4$, determine

(a) the final pressure and temperature of the gas;
(b) the external work done on the gas during compression;
(c) the heat flow which takes place.

(Ans. 242 kN/m^2; 79·5°C; $-11\cdot72$ kJ; $-6\cdot45$ kJ)

15. The bore of a gas engine is 350 mm, the stroke 405 mm, and the clearance volume 6·2 dm^3. When the piston is at inner dead centre the gas pressure is 1·92 MN/m^2, and the temperature 1100°C. The gas then expands according to the law $pV^{1\cdot35} =$ constant as the piston moves to outer dead centre. Find the work done during the expansion, the average force on the piston during the stroke, and the heat flow which occurs during the expansion, stating its direction.
Take $c_v = 710$, $R = 290$ J/kg degC.

(Ans. 17·05 kJ, 42·1 kN, 2·46 kJ)

16. In the characteristic equation $pV = RT$ for unit mass of gas, prove that R is equal to the difference between the specific heats at constant pressure and constant volume. A quantity of nitrogen gas weighs 0·2 kg and is at 15°C. It is compressed adiabatically to one-quarter of its original volume, its temperature rising to 230°C, and 32·2 kJ of work are expended on the gas during the compression. Calculate the specific heats at constant pressure and constant volume and the gas constant R.

(Ans. 1050 J/kg degC; 749 J/kg degC; 299 J/kg degC)
[U.E.I.]

17. Define the expression 'adiabatic change' as applied to a gas. Derive an expression for the work done during the ideal adiabatic expansion of a gas. 1 kg of a certain gas at a pressure of 105 kN/m^2 and a temperature of 15°C are compressed adiabatically to a pressure of 840 kN/m^2. Determine the work done during the compression. For the gas the specific heats at constant pressure and constant volume are 815 J/kg degC and 628 J/kg degC respectively.

(Ans. $-110\cdot5$ kJ)
[U.L.C.I.]

18. Show that if a quantity of gas expands according to the law $pV^n = c$, then the heat supplied during the expansion is given by:—

$$Q = \frac{\gamma - n}{\gamma - 1} \times W$$

where γ is the adiabatic index, and W is the work done.

93

A cylinder fitted with a piston contains 60 dm³ of air at 1·4 MN/m² and 150°C. The air is allowed to expand to 180 dm³, the law of expansion being $pV^{1\cdot2}=$ constant.

Calculate the final temperature of the air and the amount of heat flow through the cylinder walls, stating the direction of flow.

For air $\gamma = 1\cdot4$.

(Ans. 66·5°C; 41·03 kJ received)
[U.E.I.]

19. State what you understand by:—

    (a) an isothermal operation;
    (b) an adiabatic operation.

If a perfect gas expands adiabatically, show that the index of expansion is given by the ratio c_p/c_v where c_p and c_v are the specific heats of the gas at constant pressure and constant volume respectively.

[U.E.I.]

20. 0·5 kg of air is first expanded isothermally at a temperature of 235°C from 3·5 MN/m² to 2·1 MN/m² and then adiabatically to 140 kN/m². The air is then cooled at constant pressure and is finally restored to its initial state by adiabatic compression. Calculate the external work done by the air per cycle and find the thermal efficiency of the cycle.

For air take $\gamma = 1\cdot4$ and $c_p = 1$ kJ/kg degC.

(Ans. 21·27 kJ; 0·574)
[U.E.I.]

21. A gas-engine cylinder has a swept volume of 30 dm³, and clearance volume of 6 dm³. At the beginning of the compression stroke the cylinder is full of a gas-air mixture weighing 0·035 kg at 105 kN/m² and 40°C. It is compressed into the clearance space according to the law $pV^{1\cdot3}=$ constant.

Calculate the work done on the mixture, its change of internal energy and the heat supplied to or rejected from the mixture, during the compression stroke.

For the mixture $c_v = 754$ J/kg degC.

(Ans. −9 kJ; 6·04 kJ; 2·96 kJ rejected)
[U.L.C.I.]

22. 0·45 kg of a gas is heated first at constant volume and then at constant pressure, so that during each stage the temperature is increased by the same amount. The initial pressure of the gas is 1·5 MN/m² and the initial temperature is 130°C. The final pressure is 3·75 MN/m². Determine the volume and tempera-

ture at (a) the end of constant volume heating; (b) the end of constant pressure heating.

Determine also the heat supplied, work done and change of internal energy, during each stage.

For the gas, $c_p = 1 \cdot 047$ kJ/kg degC, $c_v = 754$ J/kg degC.

(Ans. (a) $0 \cdot 0354$ m³, 735°C; (b) $0 \cdot 0567$ m³, 1340°C. $Q = 205 \cdot 3$ kJ, $W = 0$, $E = 205 \cdot 3$ kJ; $Q = 285$ kJ, $W = 80$ kJ, $E = 205 \cdot 3$ kJ)

[U.L.C.I.]

23. Air entering the cylinder of a compressor is at a pressure of 105 kN/m² and temperature 21°C. Compréssion takes place according to $pV^n = $ constant, until the pressure is $0 \cdot 9$ MN/m². If the final volume is one-sixth of the initial volume, determine (a) the value of the index n; (b) the work done in compression, per kg of air; (c) the change of internal energy of 1 kg of air; (d) the heat supplied to or rejected from the cylinder contents during the compression stroke, per kilogramme of air.

For air, $R = 287$ J/kg degC, $cv = 718$ J/kg degC.

(Ans. $1 \cdot 2$, 181 kJ; $90 \cdot 5$ kJ; $90 \cdot 5$ kJ rejected)

[U.L.C.I.]

24. Define the terms isothermal and adiabatic as applied to certain changes in state of a perfect gas.

1 kg of a gas at 180°C is expanded adiabatically to three times its original volume. If the specific heats at constant pressure and constant volume are 990 J/kg degC and 703 J/kg degC respectively, determine (a) the final temperature of the gas; (b) the work done by the gas during expansion.

(Ans. $16 \cdot 4$°C; $115 \cdot 1$ kJ)

[U.L.C.I.]

25. State the general energy equation for a flow process explaining the meaning of each term.

A rotary air compressor increases the pressure of the air from 70 kN/m² to 125 kN/m² and the volume of the air reduces from $0 \cdot 7$ m³/kg to $0 \cdot 465$ m³/kg.

The internal energy is increased by $39 \cdot 5$ kJ/kg and $55 \cdot 8$ kJ/kg of work is done on each kilogramme of air. Find the change of enthalpy of 7 kg of air and the heat taken in or rejected by this air during its passage through the compressor. The change of potential and velocity energies may be neglected.

(Ans. $340 \cdot 4$ kJ increase, $50 \cdot 23$ kJ rejected)

26. 1 kg of air is expanded according to the law $pV^n = $ constant from initial conditions of $1 \cdot 3$ MN/m² and 250°C until it occupies $1 \cdot 25$ m³ at 37°C. Calculate the final pressure of the air and the value of the index n. Determine also the work done by the air,

and the heat flow into or out of the gas during the expansion process.

For air, $\gamma = 1.4$ and $R = 287$ J/kg degC.

(Ans. 71·2 kN/m², 1·22, 277·6 kJ; 125·1 kJ)

[U.L.C.I.]

27. 1 kg of a gas at 3 MN/m² and 2115°C is expanded to a pressure of 200 kN/m² according to the law $pV^{1·32} = $ constant.

Determine (a) the final temperature; (b) the work done by the gas; (c) the change of internal energy of the gas; (d) the heat interchange between the gas and the cylinder walls, stating whether this is added to the gas or subtracted from the gas.

For the gas $\gamma = 1.41$ and $c_p = 1$ kJ/kg degC.

(Ans. 966°C; 1·044 MJ; −814 kJ; 229 kJ)

[U.E.I.]

28. Derive an expression for the work done by a gas when it is expanded adiabatically from conditions p_1 and V_1 to p_2 and V_2.

A quantity of air occupies 0·1 m³ at 1·83 MN/m² and 330°C. After expansion to 100 kN/m² the temperature of the air has fallen to 10°C. Determine the index of expansion and compare it with the adiabatic index for air. How much heat was absorbed by the air from its surroundings during this operation?

Take $c_p = 1$ kJ/kg degC and $c_v = 708$ J/kg degC, for air.

(Ans. 1·351; 1·413; 41·53 kJ)

[U.E.I.]

29. Derive an expression for the work done by a gas when it is expanded isothermally through a volume ratio r.

0·15 m³ of air at 1·1 MN/m² and 77°C are expanded isothermally until the volume is 0·75 m³. Determine (a) the weight of air present; (b) its final pressure, and (c) the work done by the air during expansion. How much heat has been absorbed by the air from its surroundings during this expansion?

For air $R = 287$ J/kg degC.

(Ans. 1·643 kg; 220 kN/m²; 265·5 kJ; 265·5 kJ)

[U.L.C.I.]

30. Prove the relationship $c_p - c_v = R$ for a perfect gas, where c_p and c_v are the specific heats at constant pressure and constant volume respectively, and R is the characteristic constant for the gas.

1 kg of a gas is heated first at constant volume and then at constant pressure so that during each stage the same quantity of heat is supplied to the gas. If the gas is initially at 400 kN/m² and 50°C, and the final pressure is 1·2 MN/m², determine the

volume and temperature at (a) the end of constant volume heating and (b) the end of constant pressure heating.

For the gas $\gamma = 1\cdot33$ and $R = 250$ J/kg degC.

(Ans. $0\cdot2018$ m³; 696°C; $0\cdot303$ m³; 1182°C)

[U.L.C.I.]

31. A certain quantity of air occupies 34 dm³ at 850 kN/m² and 210°C. It is expanded to 100 kN/m² according to the law $pV^{1\cdot25}$ constant.

Determine (a) the final volume and temperature; (b) the work done by the air; (c) the heat absorbed by the air during this expansion, and (d) the change of internal energy of the air.

For air $\gamma = 1\cdot4$ and $R = 287$ J/kg degC.

(Ans. $0\cdot1659$ m³, $4\cdot3$°C; $35\cdot17$ kJ; $4\cdot4$ kJ; $-30\cdot77$ kJ)

[U.L.C.I.]

32. In a thermodynamic steady-flow system it is known that the internal energy per kg of fluid decreases by 92 kJ/kg. At the same time the flow-work increases from inlet to outlet by $1\cdot1$ MJ for a total flow of 25 kg. Calculate the change in specific enthalpy.

If the system is horizontal and there is no change in velocity, calculate the power produced when the flow is 50 kg/min and there is a heat flow from the system of 7 kJ/kg.

[Ans. 48 kJ/kg; $34\cdot17$ kW)

33. A perfect gas flows steadily through a cooler. The entry temperature and velocity are 315°C and 150 m/s respectively. The gas leaves the cooler through a cross-sectional area of $0\cdot2$ m² at a pressure of 250 kN/m² and at a temperature of 37°C. The rate of flow of the gas is $13\cdot5$ kg/s, it has a characteristic constant of 297 J/kg degC and its specific heat at constant pressure is $1\cdot035$ kJ/kg degC. Calculate the rate of heat transfer from the gas.

(Ans. $2\cdot907$ MW)

[U.L.C.I.]

97

5. Cycles of Operation

The following table gives average results of overall thermal efficiency for a variety of engines:—

Diesel Engines	36%
Petrol Engines	28%
Steam Turbines	30%

Reciprocating Steam Engines:—

Industrial Condensing Engines	10%
Express Locomotives	8%
Shunting Locomotives	4%

These figures show that even with the most efficient engines, only about one-third of the heat value of the fuel is converted into work, the remaining two-thirds being distributed principally to exhaust and cooling.

This appears, at first sight, to be a most unsatisfactory state of affairs, and we must now find out why these efficiencies are so low and whether they are the best that can be obtained.

The first question to be answered is:—

'Given a source of heat, what is the best way of converting it into mechanical work?' The correct answer was first given by Sadi Carnot in an essay written in 1824. In this he proved that the heat had to pass through an engine working on a cycle composed entirely of reversible operations.

REVERSIBLE OPERATIONS

An expansion is termed 'reversible' if it is carried out in such a way that, at the end of the expansion, the working agent could be compressed and pass through precisely the same path on the pV

98

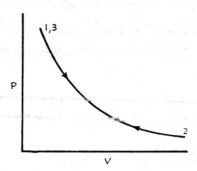

Fig. 42. Reversible Operations

diagram, causing exactly as much heat to flow back to the source as came from it during the expansion.

This means:—

(1) The expansion and contraction of the working agent must be smooth and free from eddies or turbulence, which represent irrecoverable degradation of energy. Such effects would mean that more work would need to be expanded on the compression than was obtained from the expansion.

(2) No friction must occur between moving parts.

(3) If heat is received from a source and, during the reversed action, returned to the source, then the operation can only be reversible if the heat source and the working agent are at the same temperature. This follows from the Second Law of Thermodynamics which says that heat cannot flow from a lower to a higher temperature without the aid of external work.

It follows from these requirements that a *practical* reversed action is not possible, but we shall see that the more nearly reversible any process can be made, the more work can be obtained from a given amount of energy.

Two of the expansions already discussed are, however, theoretically reversible. These are:—

(a) **Adiabatic Expansion:** Since no heat flow takes place, if this operation is carried out without frictional effects it is reversible.

(b) **Isothermal Expansion:** Since heat flow takes place at constant temperature, this operation is reversible if frictional effects are avoided.

99

THE CARNOT CYCLE OF OPERATIONS

Carnot devised the following cycle consisting entirely of reversible operations.

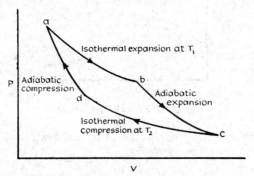

Fig. 43. Carnot Cycle

(*a–b*) *Isothermal expansion* taking in heat at the temperature of the heat source T_1.

(*b–c*) *Adiabatic expansion* during which the temperature falls from T_1 to that of a heat receiver (or sink) T_2.

(*c–d*) *Isothermal compression* rejecting heat at the temperature of the heat receiver T_2.

(*d–a*) *Adiabatic compression* during which the temperature is raised from T_2 to the heat source temperature T_1.

The net work done by such an engine would be represented by the enclosed area *abcd*.

Now such an engine, since it consists only of reversible actions, could operate in the reverse direction so that it follows the cycle shown in Fig. 44. Heat is being taken in at the lower temperature T_2

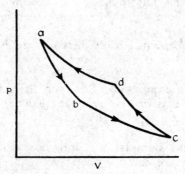

Fig. 44. Carnot Cycle Reversed

100

during the isothermal expansion (*b–c*) and is being rejected at the upper temperature T_1. Of course it will now need to be *driven*, and the enclosed area represents the work needing to be done *on* it, being exactly the same quantity as that done by it when used as an engine. When working in this way such an engine is called a *heat pump*, because it pumps up energy from a lower temperature for delivery at a higher temperature.

We can now show that such an engine, working entirely on reversible operations, has the highest possible efficiency that can be obtained. Suppose a Carnot engine were used to drive a Carnot heat pump as shown in Fig. 45.

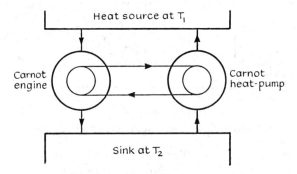

Fig. 45. Theoretical Diagram for Highest-efficiency Engine

The engine would take in heat at temperature T_1, do work equal to area *abcd*, and reject the heat it could not use to the sink at temperature T_2. The work done would be exactly the amount required to drive the heat pump which would take in heat from the sink at T_2 (the same amount as the engine rejects) and reject heat at the source at T_1 (the same amount as the engine takes in). This arrangement gives perpetual motion, but remember that so far the argument is a purely theoretical one!

Now assume there *is* an engine working on a cycle supposed to be more efficient than the Carnot engine. If it is used to drive a Carnot heat pump then it will need *less* heat than the Carnot engine to produce the work needed to drive the heat pump. This would mean that the combination would be returning more heat to the source at the higher temperature T_1 than is being taken out. But this would directly contravene the Second Law of Thermodynamics which says that heat cannot flow from a lower to a higher temperature without the aid of work from an outside source, and so we conclude that no such more efficient engine exists.

We have now achieved a very important step in the investigation

101

into engine efficiencies, because we have identified an engine with the best possible efficiency, i.e. one consisting only of reversible operations. The next step is to find out what is the efficiency of the Carnot engine which we now know to be the highest that can possibly be attained.

Thermal Efficiency of an Engine Working on the Carnot Cycle

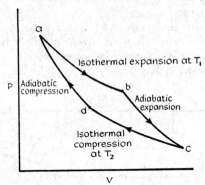

Fig. 46. Thermal Efficiency

The general expression for thermal efficiency of an engine may be written

$$\eta_t = \frac{\text{Work done per cycle}}{\text{Heat taken in per cycle}}$$

Now from the First Law for a closed system operating a closed cycle, the work done is the difference between the heat taken in and the heat rejected

$$\text{i.e. } \eta_t = \frac{\text{Heat taken in/cycle} - \text{Heat rejected/cycle}}{\text{Heat taken in/cycle}}$$

$$= 1 - \frac{\text{Heat rejected/cycle}}{\text{Heat taken in/cycle}}$$

Since the actions b–c and d–a are adiabatics during which heat is neither taken in nor rejected, the heat taken in (during the isothermal expansion a–b) is given by:—

$$Q_1 = W + (U_2 - U_1)$$

$$= p_a V_a \log_e \frac{V_b}{V_a} + 0$$

Heat rejected (during the isothermal compression c–d)

$$Q_2 = p_c V_c \log_e \frac{V_c}{V_d} + 0$$

$$\therefore \ \eta_t = 1 - \frac{p_c V_c \log_e \dfrac{V_c}{V_d}}{p_a V_a \log_e \dfrac{V_b}{V_a}}$$

To simplify this expression, we may write

$$p_a V_a = mRT_a = mRT_1$$
$$p_c V_c = mRT_c = mRT_2$$

For the adiabatic expansion b–c

$$\frac{T_2}{T} = \left(\frac{V_b}{V_c}\right)^{\gamma-1}$$

i.e. $\dfrac{V_b}{V_c} = \left(\dfrac{T_2}{T_1}\right)^{\frac{1}{\gamma-1}}$

For the adiabatic compression d–a

$$\frac{T_2}{T_1} = \left(\frac{V_a}{V_d}\right)^{\gamma-1}$$

i.e. $\dfrac{V_a}{V_d} = \left(\dfrac{T_2}{T_1}\right)^{\frac{1}{\gamma-1}}$

Hence $\dfrac{V_b}{V_c} = \dfrac{V_a}{V_d}$ or, rearranging, $\dfrac{V_b}{V_a} = \dfrac{V_c}{V_d}$

Taking logs to base e, $\log_e \dfrac{V_b}{V_a} = \log_e \dfrac{V_c}{V_d}$

$$\therefore \quad \eta_t = 1 - \frac{mRT_2 \log_e \dfrac{V_c}{V_d}}{mRT_1 \log_e \dfrac{V_b}{V_a}}$$

or $\eta_t = 1 - \dfrac{T_2}{T_1}$

Practical Lessons from the Carnot Cycle

An engine designer will strive for the best efficiency that he can obtain, with due regard to cost and manufacturing problems, for two good reasons:—

1. The engine will be less costly to run.
2. The more of the heat taken in that can be converted to useful work, the less there remains to be rejected and cause trouble. Most of the difficulties occurring in engine running arise from the heat that cannot be used and must be rejected.

Many attempts have been made to operate an engine on the Carnot cycle without much success. The principal difficulty is that a very high maximum pressure and a very long stroke are required to give a worthwhile amount of work, so that the engine becomes very heavy and the consequent friction absorbs the greater part of the work done. Nevertheless, the fact that the Carnot efficiency cannot be exceeded makes it most useful for comparison and leads to the following conclusions which are applicable to the design of any engine:—

1. Since efficiency increases with T_1 and decreases with T_2, the aim should always be to take in heat at the highest possible temperature and reject all that cannot be used at the lowest possible temperature.

2. The sink to which the heat is rejected is normally the atmosphere which has an average temperature of about 15°C. Hence for all practical purposes, the maximum possible efficiency becomes

$\eta_t = 1 - \dfrac{288}{T_1}$. Thus, for an engine taking in heat at say 880°C,

$$\text{Maximum } \eta_t = 1 - \frac{288}{1153}$$

$$= 0 \cdot 75$$

The real efficiency would be very much less than this, of course, because of unavoidable mechanical and fluid friction. In other words, *there is a limit* to the efficiency at which an engine can operate.

THE CONSTANT VOLUME OR OTTO CYCLE

In petrol and gas engines, the heat is taken in very rapidly from the combustion of the fuel whilst there is negligible movement of the piston at the top dead centre position. Similarly, the heat is rejected rapidly from the cylinder when the exhaust valve opens at or about bottom dead centre. Thus the heat exchange operations occur whilst the volume of the gases in the cylinder remains constant. The first engine to be operated successfully in this way was invented in 1876 by a German engineer named Otto. The ideal cycle corresponding to this arrangement is called the *Constant Volume Cycle*, and is illustrated in Fig. 47. For simplicity, the working agent is assumed to be air in a cylinder which receives heat perfectly when the piston is at inner dead centre and rejects heat perfectly when the piston is at outer dead centre.

The cycle consists of

a–b Adiabatic compression, so that no heat flow takes place, whilst the pressure increases from p_a to p_b, the temperature increases from T_a to T_b, and the volume decreases from V_a to V_b (= the clearance volume).

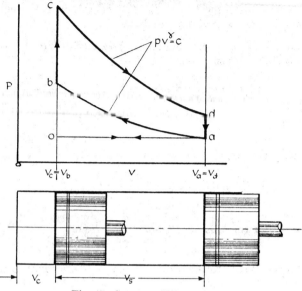

Fig. 47. Constant Volume Cycle

b–c Heat is taken in at constant volume whilst the pressure and temperature increase from p_b and T_b to p_c and T_c respectively.

c–d Adiabatic expansion, whilst the air expands from volume V_c to V_d and does work on the piston. The pressure and temperature fall from p_c and T_c to p_d and T_d respectively, and no heat is taken in or given out during the process.

d–a Heat is rejected at constant volume as the pressure and temperature fall to the original values of p_a and T_a.

(The addition of the lines a–o and o–a show the cycle for four-stroke operation, but the corresponding areas below these lines, representing exhaust and suction strokes, cancel each other out and have no effect in the ideal case.)

We can now obtain the ideal or 'air standard' efficiency, as it is called, for an engine working on such a cycle.

$$\text{Air standard efficiency} = \frac{\text{Heat taken in/cycle} - \text{Heat rejected/cycle}}{\text{Heat taken in/cycle}}$$

$$= 1 - \frac{\text{Heat rejected/cycle}}{\text{Heat taken in/cycle}}$$

The expansion and compression operations are both adiabatic in which there is no heat flow across the boundaries. Remembering that the heat required to raise m kg of gas from temperature T_1 to

105

temperature T_2 is given by $m \times c_v \times (T_2 - T_1)$ where $c_v =$ specific heat capacity at constant volume, we have

heat taken in at constant volume $(b-c) = mc_v(T_c - T_b)$
heat rejected at constant volume $(d-a) = mc_v(T_d - T_a)$

$$\therefore \text{ air standard efficiency} = 1 - \frac{mc_v(T_d - T_a)}{mc_v(T_c - T_b)}$$

$$= 1 - \frac{(T_d - T_a)}{(T_c - T_b)}$$

To simplify this expression, we may write:—

$$\text{for the adiabatic } (a - b) \quad \frac{T_a}{T_b} = \left(\frac{V_b}{V_a}\right)^{\gamma-1}$$

$$\text{for the adiabatic } (c - d) \quad \frac{T_d}{T_c} = \left(\frac{V_c}{V_d}\right)^{\gamma-1}$$

Now observe that both the ratios $\dfrac{V_a}{V_b}$ and $\dfrac{V_d}{V_c}$ are equal to the compression ratio r, for $r = \dfrac{V_c + V_s}{V_c}$

$$\therefore \ T_a = T_b\left(\frac{1}{r}\right)^{\gamma-1} \qquad \text{and } T_d = T_c\left(\frac{1}{r}\right)^{\gamma-1}$$

$\therefore$ Substituting for T_d and T_a

$$\text{Air standard efficiency} = 1 - \frac{\left(T_c\left(\frac{1}{r}\right)^{\gamma-1} - T_b\left(\frac{1}{r}\right)^{\gamma-1}\right)}{T_c - T_b}$$

$$\therefore \text{ Air standard efficiency} = 1 - \frac{1}{r^{\gamma-1}}\frac{[T_c - T_b]}{[T_c - T_b]}$$

$$\text{A.S.E.} = 1 - \frac{1}{r^{\gamma-1}}$$

The Constant Volume Cycle in Practice

The shape of the actual indicator diagram traced out by an engine in practice follows the ideal diagram fairly closely. The action of a practical engine differs in many respects, however, from the ideal cycle described because:—

1. The heat is released from within the working agent which is a combustible mixture and not air alone.

2. Complicated exchanges of heat take place between the cylinder walls and the cooling jackets throughout the cycle.

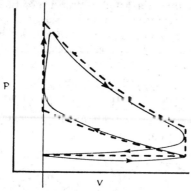

Fig. 48. Constant Volume Cycle in Practice

3. The heat from the charge takes an appreciable time to be released so that the heat intake is not exactly at constant volume. Similarly, the exhaust valve takes some time to open as is shown by the rounding-off of the diagram at this point, so that the heat rejection also does not take place exactly at constant volume.

Nevertheless, the expression for air standard efficiency gives a most useful point of comparison for all engines working on this cycle. The first point to notice is that the thermal efficiency increases as the compression ratio increases, and this important fact is borne out in practice.

The graph in Fig. 49 shows how the air standard efficiency changes with compression ratio r.

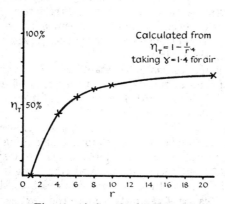

Fig. 49. Air Standard Efficiency

In the design of internal combustion engines, every effort is made to use as high a compression ratio as possible. Unfortunately, with

107

petrol and gas engines, in which a mixture of fuel and air is compressed into the clearance space during the compression stroke, the maximum compression ratio which can be used is about 10 to 1 owing to a phenomenon known as 'detonation' or 'pinking'.

DETONATION

In order to understand what is meant by 'detonation' let us first examine what happens in an engine cylinder from the instant the spark occurs at the plug points.

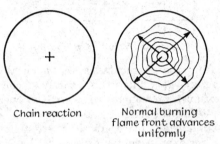

Chain reaction   Normal burning
flame front advances
uniformly

Fig. 50. Diagram of Normal Burning

The first stage lasts for about $\frac{1}{1000}$ of a sec., and is known as the ignition delay period. This time elapses before the flame first appears and is believed to be due to the setting up of a chain reaction within the molecules of the mixture before they are ready to burn. It is this ignition delay period which makes it necessary to alter the instant of sparking so as to ensure that heat release and the resulting pressure rise always occur at inner dead centre whatever the engine speed. This alteration of the instant of sparking is known as 'advancing or retarding the ignition'.

In the next stage, the flame is established as a ragged ring which proceeds to spread outwards at a steady rate of about 15 m/s. As the flame spreads outwards heat is released from within the ring with a consequent rise in pressure.

Under normal conditions, the flame spread proceeds uniformly throughout the clearance space which forms the combustion chamber, releasing heat as it goes. However, a fuel mixture may be made to ignite in two different ways. One method is to bring it into intimate contact with a flame, so that the section nearest the flame becomes heated and ignites, passing on its heat to the next section which in turn ignites, and so on. In this way the burning proceeds smoothly and the heat release occurs at a uniform rate. A second method consists of heating up the whole mixture to its ignition temperature at

108

which point the whole mixture fires spontaneously so that the heat release occurs very suddenly.

Now it may happen that, during the burning process in the combustion chamber, while the flame spreads outwards, the rising pressure resulting from the heat release may so compress that part of fuel as yet unburnt that it fires spontaneously.

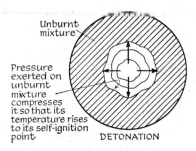

Unburnt mixture

Pressure exerted on unburnt mixture compresses it so that its temperature rises to its self-ignition point

DETONATION

Fig. 51. Detonation

Spontaneous ignition means that heat release from the remaining fuel occurs very suddenly with an extremely high rate of pressure increase. This is known as detonation or pinking, and has most serious effects on the engine. The high-pressure wave produced gives a violent blow to the walls of the combustion chamber and the piston crown, producing a sound like a hammer striking metal. Detonation is very likely to cause material damage and always results in over-heating, so that it must be avoided at all costs.

It should be noticed that as the compression ratio is increased, the initial pressure of the fuel mixture before burning is increased so that detonation becomes more likely. Many other factors influence the onset of detonation, including the design of the combustion chamber and the quality of the fuel, but the limiting factor is that of compression ratio.*

THE SLOW-SPEED DIESEL CYCLE

The essential difference between the diesel engine and petrol or gas engines is that *air alone* is compressed into the clearance space, the fuel being injected into the cylinder after compression when the air temperature is sufficient to ignite it (with slow-speed diesel engines, the oil fuel is forced into the cylinder by a blast of compressed air). The fuel is ignited almost immediately by the hot air

* If you want to know more about detonation, read Ricardo's *The High-Speed Internal-Combustion Engine.*

in the cylinder and heat release takes place at a rate sufficient to maintain the pressure in the cylinder constant as the piston is driven out for part of its stroke. After the fuel is cut off, expansion of the hot gases takes place until the exhaust ports are uncovered at the end of the stroke. The ideal cycle for a slow-speed diesel engine is as shown in Fig. 52.

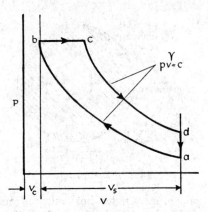

Fig. 52. Slow-speed Diesel Cycle

(a–b) Adiabatic compression—no heat flow.
(b–c) Heat intake at constant pressure, the heat supply ceasing at point c which is called the point of cut-off.
(c–d) Adiabatic expansion—no heat flow.
(d–a) Heat rejection at constant volume.
   We can now obtain the Ideal efficiency for this cycle.

$$\text{Ideal cycle efficiency} = \frac{\text{Heat taken in/cycle} - \text{heat rejected/cycle}}{\text{Heat taken in/cycle}}$$

$$= 1 - \frac{\text{Heat rejected/cycle}}{\text{Heat taken in/cycle}}$$

During the adiabatic operations no heat is taken in or rejected, hence

Heat taken in (at constant pressure b–c) $= mc_p(T_c - T_b)$
Heat rejected (at constant volume d–a) $= mc_v(T_d - T_a)$

$$\therefore \text{Ideal cycle efficiency} = 1 - \frac{mc_v(T_d - T_a)}{mc_p(T_c - T_b)}$$

$$= 1 - \frac{T_d - T_a}{\gamma\,(T_c - T_b)}$$

The Slow-speed Diesel Engine Cycle in Practice

Slow-speed diesel engines are mainly used as marine engines, usually working on the two-stroke cycle, the upper pressure being of the order of 4 MN/m² (and running speeds up to 250 rev/min). Since air alone is contained in the cylinder during the compression stroke, the problem of detonation does not arise. The expression obtained for ideal cycle efficiency can be developed as shown in the footnote below* and again, as in the case of the Otto cycle, the efficiency increases with increased compression ratio. Now, since detonation is no longer a problem, high compression ratios can be used and values of r range from 15 to 19. As a result the diesel engine has a higher thermal efficiency than a corresponding petrol engine.

THE HIGH-SPEED DIESEL CYCLE

In any diesel engine running at more than, say, 350 rev/min, it is necessary to start injection of the fuel oil before the air temperature is sufficient to ignite it. This is because:—

(i) The operation of injecting the oil takes time, and with high piston speeds early injection is essential to deliver the required quantity.

(ii) The injected oil is at atmospheric temperature and has first to be warmed up by the air in the cylinder before it ignites.

* From ideal cycle efficiency $= 1 - \dfrac{T_d - T_a}{\gamma(T_c - T_b)}$ obtain all temperatures in terms of the temperature at the end of compression T_b.

For adiabatic a–b $\quad \dfrac{T_a}{T_b} = \left(\dfrac{V_b}{V_a}\right)^{\gamma-1} \quad \therefore T_a = T_b\,\dfrac{1}{r^{\gamma-1}}$

For heat intake b–c, let $\dfrac{V_c}{V_b} = \rho$, the cut-off ratio

$$\dfrac{p_c V_c}{T_c} = \dfrac{p_b V_b}{T_b} \quad \therefore T_c = T_b \times \dfrac{V_c}{V_b} = \rho\,.\,T_b\,.$$

For adiabatic c–d $\quad \dfrac{T_d}{T_c} = \left(\dfrac{V_c}{V_d}\right)^{\gamma-1} = \left(\dfrac{V_c}{V_b}\,.\,\dfrac{V_b}{V_c}\right)^{\gamma-1} = \dfrac{\rho\,.\,\gamma-1}{r^{\gamma-1}}$

$$\therefore T_d = T_c\,\dfrac{\rho^{\gamma-1}}{r^{\gamma-1}} = T_b\,\dfrac{\rho^\gamma}{r^{\gamma-1}}$$

$\therefore$ Substituting for temperature

$$\text{Ideal cycle efficiency} = 1 - \dfrac{T_b\dfrac{\rho^\gamma}{r^{\gamma-1}} - T_b\dfrac{1}{r^{\gamma-1}}}{\gamma\,[\rho T_b - T_b]}$$

$$\text{Ideal cycle efficiency} = 1 - \dfrac{1}{r^{\gamma-1}}\,.\,\dfrac{\rho^\gamma - 1}{\gamma(\rho - 1)}$$

111

It follows that when ignition does occur, there is a very rapid rise of pressure occurring at approximately constant volume due to the heat release from the oil which has accumulated during the early injection period. The oil which is injected after ignition burns at approximately constant pressure.

The ideal cycle for this type of engine is the *Dual-combustion Cycle* in which the intake of heat is partly at constant volume and partly at constant pressure as shown in Fig. 53.

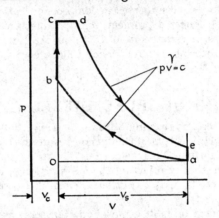

Fig. 53. High-speed Diesel Cycle

An expression for the ideal efficiency of such a cycle can be obtained, but in most modern high-speed diesel engines the period of constant pressure burning is so short that the cycle approximates very closely to the constant volume or Otto cycle. Accordingly it has been recommended that the performance of all types of internal combustion engines be referred to the air standard efficiency of the constant volume cycle, namely

$$\text{A.S.E.} = 1 - \frac{1}{r^{\gamma-1}}$$

RELATIVE EFFICIENCY

This is the relationship between the actual and ideal thermal efficiencies of an engine. The actual thermal efficiency used should be that based on i.h.p., thus the relative efficiency of an engine operating on the constant volume cycle becomes

$$\text{Relative efficiency} = \frac{\text{Indicated thermal efficiency}}{1 - \dfrac{1}{(r)^{\gamma-1}}}$$

112

Example 22

'A single-cylinder gas engine working on the four-stroke cycle is required to develop brake power of 20 kW with a mechanical efficiency of 80% when running at 300 rev/min. The mean effective pressure is to be 600 kN/m^2 and the stroke $= 1\frac{1}{2} \times$ bore. Determine the necessary cylinder diameter.

If the clearance volume is one-quarter of the swept volume, determine the air standard efficiency. Assuming the relative efficiency as 55%, estimate the specific fuel consumption (brake basis) if the calorific value of the gas $= 18$ MJ/m^3. Take $\gamma = 1\cdot4$.'

$$\text{i.p.} = \frac{\text{b.p.}}{\text{Mech.effy}} = \frac{20}{0\cdot8} = 25 \text{ kW}$$

Let cylinder diameter $= d$ m

$$\therefore \text{ stroke} = 1\cdot5d \text{ m}$$

$$\text{i.p.} = p_m LAn$$

$$\therefore \ 25 \times 10^3 = 600 \times 10^3 \times \frac{\pi d^2}{4} \times 1\cdot5d \times \frac{300}{2 \times 60}$$

$$\therefore \ d = \sqrt[3]{\frac{25 \times 10^3 \times 4 \times 2 \times 60}{600 \times 10^3 \times \pi \times 1\cdot5 \times 300}}$$

$$= 0\cdot2418 \text{ m}$$

Compression ratio $r = \dfrac{V_c + V_s}{V_c}$

$$= \frac{0\cdot25 \ V_s + V_s}{0\cdot25 \ V_s}$$

$$= 5$$

$\therefore$ air standard efficiency $= 1 - \dfrac{1}{r^{0\cdot4}}$

$$= 1 - \frac{1}{5^{0\cdot4}}$$

$$= 0\cdot474$$

since relative efficiency $= 0\cdot55$

$\therefore$ indicated thermal efficiency $= 0\cdot55 \times 0\cdot474$
$$= 0\cdot261$$

$\therefore$ brake thermal efficiency $\quad = 0\cdot261 \times 0\cdot8$
$$= 0\cdot2085$$

$$\text{Brake thermal efficiency} = \frac{\text{b.p.} \times 60}{\text{m}^3 \text{ of fuel/min} \times \text{c.v.}}$$

$$= \frac{\text{b.p.} \times 3600}{\text{m}^3 \text{ of fuel/hour} \times \text{c.v.}}$$

$\therefore$ Specific fuel consumption (m³ kWh) (brake basis)

$$= \frac{\text{m}^3 \text{ of fuel/hour}}{\text{b.p. (kW)}}$$

$$= \frac{3600 \times 10^3}{\text{Brake thermal effy} \times \text{c.v.}}$$

$$= \frac{3600 \times 10^3}{0 \cdot 2085 \times 18 \times 10^6}$$

$$= \underline{0 \cdot 959 \text{ m}^3/\text{kWh}}$$

Example 23

'A gas engine bore 250 mm, stroke 460 mm has a compression ratio of $4\frac{1}{2}$. At the end of suction the pressure is 90 kN/m². Compression equation $pV^{1\cdot35} = c$. Expansion equation $pV^{1\cdot3} = c$. If the pressure is trebled during the constant volume explosion, find the mean effective pressure and the i.p. developed if the engine makes 85 explosions/minute.' [U.E.I.]

$$p_2 = p_1\left(\frac{V_1}{V_2}\right)^{1\cdot35} = 90 \times 4 \cdot 5^{1\cdot35} = 685 \cdot 5 \text{ kN/m}^2$$

$$p_3 = 3p_2 = 2056 \cdot 5 \text{ kN/m}^2$$

$$p_4 = p_3\left(\frac{V_3}{V_4}\right)^{1\cdot3} = \frac{2056 \cdot 5}{(4 \cdot 5)^{1\cdot3}}$$

$$= 290 \cdot 9 \text{ kN/m}^2$$

$$P_m = \frac{\text{Area of diagram}}{\text{Base length}}$$

$$= \frac{(\text{Area under curve 3-4}) - (\text{Area under curve 2-4})}{\text{Base length}}$$

$$P_m = \frac{\dfrac{p_3V_3 - p_4V_4}{n_e - 1} - \dfrac{p_2V_2 - p_1V_1}{n_c - 1}}{V_1 - V_2}$$

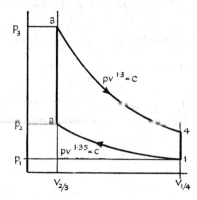

Fig. 54. Gas Engine Cycle

Now $V_1 = 4 \cdot 5\ V_2$ (and $V_4 = 4 \cdot 5\ V_3$)

$\therefore \quad P_m =$

$$\frac{10^3 \left[\dfrac{2056 V_3 - (290 \cdot 9 \times 4 \cdot 5) V_3}{0 \cdot 3} - \dfrac{685 \cdot 5\ V_2 - (90 \times 4 \cdot 5)\ V_2}{0 \cdot 35} \right] \text{N/m}^2}{3 \cdot 5\ V_2}$$

Since $V_2 = V_3$, volumes will cancel

$$\therefore \quad P_m = \left[\frac{2056 - 1309}{0 \cdot 3 \times 3 \cdot 5} - \frac{685 \cdot 5 - 405}{0 \cdot 35 \times 3 \cdot 5} \right] 10^3\ \text{N/m}^2$$

$$= 482 \cdot 3\ \text{kN/m}^2$$

$$\therefore \quad \text{i.p.} = P_m L A n$$

$$= 482 \cdot 3 \times 10^3 \times \frac{\pi}{4} \times (\cdot 25)^2 \times (\cdot 46) \times \frac{85}{60}\ \text{W}$$

$$= 15 \cdot 43\ \text{kW.}$$

Example 24

'A petrol engine and a high-speed diesel engine have the same capacity, the petrol engine using a compression ratio of 6·5 and the diesel engine 16. If, when developing the same i.p., the petrol engine has a relative efficiency of 60%, and the diesel engine a relative efficiency of 55%, both based on the Otto cycle, find the specific fuel consumption kg/kWh (indicated basis) of each engine, considering both petrol and diesel fuel to have a calorific value of 44 MJ/kg.'

For petrol engine

$$\text{Air standard efficiency} = 1 - \frac{1}{6\cdot5^{0\cdot4}} = 0\cdot528$$

$$\therefore \text{Indicated thermal efficiency} = 0\cdot528 \times 0\cdot6$$
$$= 0\cdot3168$$

$$\text{Indicated thermal efficiency} = \frac{\text{i.p.} \times 3600 \text{ J}}{\text{kg of fuel/hour} \times \text{c.v.}}$$

$\therefore$ Specific fuel consumption

$$[\text{kg/kWh (ind)}] = \frac{3\cdot6 \times 10^6}{\eta_t \times \text{c.v.}}$$

$$= \frac{3\cdot6 \times 10^6}{0\cdot3168 \times 44 \times 10^6}$$

$$= \underline{0\cdot258}$$

For diesel engine

$$\text{Air standard efficiency} = 1 - \frac{1}{16^{0\cdot4}} = 0\cdot67$$

$$\text{Indicated thermal efficiency} = 0\cdot67 \times 0\cdot55$$
$$= 0\cdot3685$$

Specific fuel consumption

$$[\text{kg/kWh (ind.)}] = \frac{3\cdot6 \times 10^6}{0\cdot3685 \times 44 \times 10^6}$$

$$= \underline{0\cdot222}$$

Example 25

'Find the ideal efficiency of a slow-speed diesel engine having a

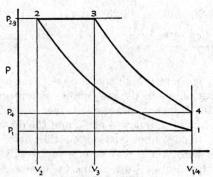

Fig. 55. Diesel Engine Cycle

compression ratio of 15, taking the temperature at the beginning of compression as 50°C, and that at the end of combustion as 1390°C.'

$$T_2 = T_1 \left(\frac{V_1}{V_2}\right)^{\gamma-1}$$

$$= 323 \times 15^{0\cdot4} = 954 \cdot 1 \text{ K}$$

$$\frac{p_2 V_2}{T_2} = \frac{p_3 V_3}{T_3} \text{ and } p_2 = p_3$$

$$\therefore \frac{V_3}{V_2} = \frac{T_3}{T_2} = \frac{1663}{954 \cdot 1} = 1 \cdot 743$$

Now $\dfrac{V_4}{V_2} = 15$

$$\therefore \frac{V_4}{V_3} = \frac{V_4}{V_2} \times \frac{V_2}{V_3} = \frac{15}{1 \cdot 743} = 8 \cdot 606$$

$$\therefore T_4 = T_3 \left(\frac{V_3}{V_4}\right)^{\gamma-1} = 1663 \times \frac{1}{(8 \cdot 606)^{0 \cdot 4}} = 703 \cdot 1 \text{ K}$$

$$\therefore \text{ ideal efficiency} = 1 - \frac{T_4 - T_1}{\gamma\,(T_3 - T_2)}$$

$$= 1 - \frac{703 \cdot 1 - 323}{1 \cdot 4(1663 - 954 \cdot 1)}$$

$$= 0 \cdot 617$$

Example 26

'The cylinder of an oil engine is to be charged with air at 96 kN/m² and 60°C, compression then taking place to $\frac{1}{14}$th of the original volume, according to $pV^{1\cdot35} = c$. Oil fuel with a mass of $\frac{1}{50}$th of that of the air is injected into the cylinder and combustion is assumed to occur at constant pressure. If the calorific value of the oil is 44 MJ/kg, determine

(a) the theoretical pressure and temperature after compression;
(b) the theoretical temperature after combustion;
(c) the theoretical point of cut-off of the fuel to the cylinder as a fraction of the stroke. (Take c_p for mixture = 990 J/kgK.)' [L.U.]

(a) $p_1 V_1^{1\cdot35} = p_2 V_2^{1\cdot35}$

$$\therefore 96 \times 10^3 \times 14^{1\cdot35} = \underline{3 \cdot 385 \text{ MN/m}^2}$$

$$T_2 = T_1 \left(\frac{V_1}{V_2}\right)^{n-1}$$

$$= 600 \times 14^{0\cdot35} = 333 \times 14^{0\cdot35}$$

$$\therefore \underline{t_2 = 838 \cdot 5 \text{ K or } 565 \cdot 5°C}$$

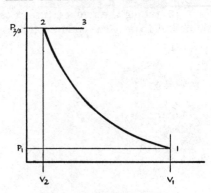

Fig. 56. Oil Engine Cycle

(b) Consider 1 kg of air. Then heat added by oil $= \frac{1}{50} \times 44 \times 10^6$
$= 880$ kJ

$$\text{Heat added} = mc_p(T_3 - T_2)$$
$$\therefore 880 \times 10^3 = 1 \cdot 02 \times 990(T_3 - 838 \cdot 5)$$
$$\therefore T_3 = 1710 \text{ K}$$
$$\therefore t_3 = 1437°\text{C}$$

(c) $\dfrac{p_3 V_3}{T_3} = \dfrac{p_2 V_2}{T_2}$ $\therefore \dfrac{V_3}{V_2} = \dfrac{T_3}{T_2} = \dfrac{1710}{838 \cdot 5} = 2 \cdot 039$

Cut-off expressed as a fraction of the stroke $= \dfrac{V_3 - V_2}{V_1 - V_2}$

$$= \dfrac{2 \cdot 039 \, V_2 - V_2}{14 V_2 - V}$$

$$= \dfrac{1}{12 \cdot 5}$$

REVISION EXERCISES—CHAPTER 5

1. The combustion chamber of a petrol engine is bounded by the flat surface of the piston crown and cylinder head, and may be considered truly hemispherical. If the compression ratio of the engine is $6\frac{1}{2}$ to 1, calculate

  (i) the stroke-bore ratio
  (ii) the air standard efficiency.

If the relative efficiency of the engine is 65% and the petrol used has a calorific value of 44 MJ/kg, calculate the specific petrol consumption in kg/kWh (ind. basis).

(Ans. 1·833, 0·527, 0·239)

2. The compression curve on the indicator diagram from a four-stroke constant-volume cycle engine follows the law $pV^{1\cdot3}=c$. At two points on the curve, at quarter stroke and three-quarter stroke, the pressures are 140 kN/m² and 364 kN/m² respectively. If the engine has a bore of 150 mm and stroke 250 mm, calculate (a) the clearance volume; (b) the compression ratio; (c) the air standard efficiency.

(Ans. 931 cm³, 5·74, 0·503)

3. The cylinder diameter and stroke of a gas engine working on the Otto cycle are 180 mm and 300 mm respectively, and clearance volume = 1900 cm³. If the relative efficiency of this engine is 50%, find the actual thermal efficiency and the gas consumption per m³/kWh, c.v. of gas being 18 MJ/m³.

(Ans. 0·2377, 0·841 m³)

4. A gas mixture is subjected to the following cycle of events in a closed system:—

(i) Adiabatic compression from a pressure of 103 kN/m² and volume 2 m³ to a volume of 0·2 m³.
(ii) Constant pressure expansion to a volume of 0·35 m³.
(iii) Adiabatic expansion back to the original volume.
(iv) Constant volume heat rejection to its original pressure.

Sketch the cycle of events on a pV diagram and determine the net work done. If the initial temperature is 10°C, determine the maximum temperature reached in the cycle. Take $c_p = 1$ kJ/kgK, $c_v = 707$ J/kgK.

(Ans. 760·2 kJ, 1001°C)

5. The compression and the expansion curves of a gas engine indicator diagram follow the law $pV^{1\cdot3}=c$. The compression begins at a pressure of 100 kN/m² and a temperature of 65°C. The temperature at the end of compression is 330°C and the maximum pressure after combustion at constant volume is 3 MN/m².

Find the pressures and temperatures at the four corners of the diagram, assuming that the exhaust valve opens at the end of the stroke. Also calculate the mean effective pressure of the indicator diagram.

	point	pressure kN/m²	temp. °C
(Ans.	1	100	65
	2	1228	330
	3	3000	1199
	4	244·2	552

m.e.p. = 441 kN/m²)

[L.U.]

6. Describe the constant-volume cycle for an air engine and calculate

the thermal efficiency of this cycle when the pressure at the
end of compression is 15 times that at the start. If, in the above
case, the initial temperature of the air at the start of compression is 37°C, and the maximum temperature reached at
the end of constant volume combustion is 1950°C, find

(a) the heat supplied/lb of air;
(b) the work done per cycle by 1 kg of air.

Take $R = 287$ J/kgK °R and $\gamma = 1.4$.

(Ans. 0·539; 1·113 MJ; 600 kJ)
[L.U.]

7. A four-cylinder petrol engine of 100 mm bore and 120 mm stroke
has a compression ratio of 6 to 1. Find the clearance volume and
the air standard efficiency. If the indicated mean effective
pressure is 650 kN/m², find the i.p. developed at 2000 rev/min.
Find also the thermal efficiency if the fuel consumption is 11·3
kg/h, the calorific value of the fuel being 43 MJ/kg.

(Ans. 188·6 cm³, 0·5119, 40·8 kW, 0·303)
[L.U.]

8. In a slow-speed diesel engine cycle the initial conditions are
100 kN/m² and 37°C. Taking the theoretical maximum pressure
and temperature to be 4·5 MN/m² and 1665°C respectively,
calculate:—

(a) the theoretical temperatures and pressures at the corners
of the diagram;
(b) the ideal thermal efficiency of the cycle.

point	pressure kN/m²	temp. °C
(Ans.　1	100	37
2	4500	646
3	4500	1665
4	284	607

thermal efficiency = 0·6)

9. The temperature of the air at the commencement of compression
in an Otto cycle is 27°C and the air receives 600 kJ/kg during
the constant volume heating. If the compression ratio is 8 : 1,
what will be the maximum temperature of the air in the cycle,
and how much heat will be rejected per kg of air during the
constant volume cooling process?

Sketch the pV diagram for this cycle, and show the pressures
at all the salient points if the minimum pressure in the cycle is
105 kN/m².

$c_v = 717$ J/kgK, $R = 287$ J/kgK.

(Ans. $t_3 = 1253$°C, $Q = 261$ kJ, $p_2 = 1.93$ MN/m²,
$p_3 = 4.27$ MN/m², $p_4 = 233$ kN/m²)
[I.Mech.E. Part I]

10. For a diesel cycle, i.e. with heating at constant pressure and cooling at constant volume, sketch the pV diagram and mark the salient points of the diagram consecutively 1, 2, 3, 4, in a clockwise direction with 1 at the commencement of compression. Show that the efficiency of the cycle is given by

$$\eta = 1 - \frac{1}{\gamma}\,\frac{(p_4 V_4 - p_1 V_1)}{(p_3 V_3 - p_2 V_2)}$$

If the compression ratio of this cycle is 12 : 1 and the expansion ratio is 8 : 1, what is the thermal efficiency using air as the working substance throughout the cycle?

(Ans. 0·60)

[I. Mech.E. Part I]

11. An Otto-cycle engine has a compression ratio of 7·0. The fuel consumption is 0·4 kg/kWh, and the calorific value of the fuel is 43 MJ/kg. What are the thermal efficiency and efficiency ratios (to the ideal air Otto cycle) of the engine?

If the brake power output is 7·5 kW, draw up a heat balance, in kJ/s, given that the friction horsepower is 2·25 kW, the cylinder cooling system takes 25 kg of water per minute with a temperature rise of 6 degC, and exhaust losses are 720 kJ/min.

(Ans. 29·9%, 0·387, Work 7·5 kJ/s, Cooling = 10·46 Friction = 2·25, not included in the account, Radiation, etc. 5·87)

[I.Mech.E. Part I]

12. Find the mean effective pressure for the ideal air standard Otto cycle, having a maximum pressure 4·2 MN/m² and minimum pressure 100 kN/m², and a compression ratio 5 : 1.

(Ans. 964 kN/m²)

[I.Mech.E. Part I]

13. Derive an expression for the mean effective pressure of the ideal Otto cycle in terms of the thermal efficiency, the heat received and the displacement. Find the m.e.p. for such a cycle in which the compression ratio is 7, the lowest temperature in the cycle is 90°C, the lowest pressure is 97 kN/m² and the temperature rise during heat reception is 1650 degC. Assume air to be the working substance throughout.

(Ans. 695 kN/m²)

[I.Mech.E. Part I]

14. Sketch the pV diagram for the ideal constant-volume cycle and the ideal diesel cycle with constant-volume heat rejection. Indicate the points of maximum and minimum temperature,

121

and hence deduce which of the two cycles will have the greater thermal efficiency for the same maximum temperature and compression ratio.

Find the thermal efficiency of the diesel cycle for a compression ratio of 15 : 1 and 'cut-off' at 10% of the cylinder volume; the working substance may be assumed to be air throughout.

(Ans. 63·05%)

[I.Mech.E. Part I]

15. What is the effect of increasing the compression ratio (a) on the efficiency; (b) on the work output per cubic metre of piston displacement, of an engine operating with the ideal Otto cycle? Find the above quantities, i.e. efficiency and work per m³ for an Otto-cycle engine working with air and compression ratio 7 : 1, the maximum temperature (abs.) in the cycle being six times the minimum. The pressure before compression may be assumed to be 100 kN/m².

(Ans. 54·2%, 600 kJ/m³)

[I.Mech.E. Part I]

16. Describe the constant volume air standard cycle and show that its thermal efficiency is given by $1 - \dfrac{1}{r^{\gamma-1}}$ where r is the volume compression ratio and γ is the adiabatic index for air.

A gas engine working on this cycle has a cylinder diameter 200 mm, a stroke 400 mm and a clearance volume 3200 cm³. It consumes 7·5 m³ of gas per hour of calorific value 18 mJ/m² when developing 10·5 kW. Compare its thermal efficiency with that of the air standard cycle.

(Ans. 28%, 47·2%)

[I.Mech.E. Part I]

17. What limits the availability of heat as a means of producing mechanical work?

The working agent in an ideal heat engine takes in heat at 600°C and rejects heat at 220°C. What is the maximum possible ideal thermal efficiency the engine could have? If the engine fuel has a calorific value of 44 mJ/kg determine the maximum theoretical work obtained per kg of fuel consumed.

Give reasons why this amount of energy conversion is not achieved in practice.

(Ans. 0·435, 19·15 MJ)

[U.E.I.]

18. In the theoretical cycle for a diesel engine, air at 105 kN/m² and 65°C is compressed adiabatically through a volume com-

pression ratio of 12. Heat is added at constant pressure until the temperature rises to 1650°C.

The air then expands adiabatically to its initial volume, when heat is rejected at constant volume such that the cycle is closed. Calculate:—

(a) the pressure and temperature at the salient points of the cycle;

(b) the external work done per kg of air.

Take $\gamma = 1.4$ and $R = 287$ J/kgK.

(Ans. (a)	kN/m^2		$°C$		
	p_1	105	t_1	65	(b) 570·5 kJ
	p_2	3405	t_2	640	
	p_3	3405	t_3	1650	
	p_4	298	t_4	686	[U.E.I.]

19. Derive an expression for the air standard efficiency of an engine working on the Otto cycle, in terms of the compression ratio 'r' and the adiabatic index 'γ'.

A gas engine has a bore and stroke of 200 mm and 250 mm respectively. The clearance volume is 1600 cm³.

Calculate the air standard efficiency and, if the relative efficiency is 60%, estimate the gas consumption in m³/kWh, taking the calorific value of the gas as 18 mJ/m³.

Take $\gamma = 1.4$.

(Ans. 0·509, 0·655)
[U.E.I.]

20. At the commencement of the compression stroke, the cylinder of an oil engine is charged with air at 100 kN/m² and at 65°C. Compression takes place to $\frac{1}{14}$ of the original volume according to the law $pV^{1.35}$ = constant. Fuel is then injected, the weight of the fuel injected being $\frac{1}{50}$ of that of the air in the cylinder, and combustion takes place at constant pressure. Taking the calorific value of the oil as 45 MJ/kg, determine:—

(a) the theoretical pressure and temperature after compression;

(b) the theoretical temperature after combustion;

(c) the fraction of the stroke at which combustion is theoretically complete.

For the mixture take $c_p = 1005$ J/kgK.

(Ans. 3·524 MN/m², 578°C, 1670°C, 1/10·1)
[U.E.I.]

21. In the ideal constant-volume cycle the compression ratio is 6·5 : 1 and the temperature at the beginning of the compression

is 37°C. Determine the air standard efficiency and the temperature at the end of compression. Assume $\gamma = 1\cdot4$.

A petrol engine develops 30 kW and uses fuel of specific
gravity 0·74 and calorific value 43 MJ/kg. If the efficiency ratio
referred to the air standard cycle is 0·60, determine the number
of cubic decimetres of fuel consumed by the engine per hour.

(Ans. 0·527, 383°C, 10·74)
[U.L.C.I.]

22. Sketch the pressure-volume diagram for the ideal diesel cycle
and state the sequence of events.

In an ideal diesel cycle, the temperatures at the beginning
and end of expansion are 1615°C and 625°C and at the beginning
and end of compression are 32°C and 615°C. Determine (a) the
volume compression ratio; (b) the percentage of the working
stroke at which combustion ceases; (c) the thermal efficiency.
Assume $\gamma = 1\cdot4$.

(Ans. 14·46, 1/10·71, 0·576)
[U.L.C.I.]

23. A single-cylinder diesel engine has a stroke and bore of 300 mm
and 250 mm respectively. The compression ratio is 14 : 1 and
at the beginning of the compression stroke the cylinder is filled
with air at 100 kN/m² and 90°C. If the compression follows
the law $pV^{1\cdot3} = $ constant, determine:—

(a) the mass of air present;
(b) the work done during compression;
(c) the heat exchange between the air and the cylinder walls
during the compression stroke, stating the direction of
flow (for air $\gamma = 1\cdot4$ and $c_p = 1007$ J/kgK).

(Ans. (a) 0·0152 kg; (b) 6·375 kJ; (c) −1·594 kJ rejected)
[U.E.I.]

24. Sketch the pressure-volume diagram for the ideal diesel cycle
and derive an expression for the efficiency of this cycle, in terms
of the temperatures at the four salient points.

For the above cycle the conditions at the beginning of compression are: pressure 100 kN/m², temperature 37°C. The
maximum temperature in the cycle is 1600°C and the volume
compression ratio is 14 : 1. Determine (a) the maximum pressure in the cycle; (b) the point in the expansion stroke at which
combustion ceases; (c) the thermal efficiency of the cycle.
Take $\gamma = 1\cdot4$ and $R = 287$ J/kgK.

(Ans. 4·023 MN/m², 8·48% of stroke, 0·587)
[U.L.C.I.]

25. Define the expressions 'Air Standard Efficiency' and 'Efficiency Ratio' as applied to an internal combustion engine.

A six-cylinder four-stroke petrol engine of cylinder bore 90 mm, stroke 100 mm, has a volume compression ratio of 7 : 1. The relative efficiency, referred to the air standard cycle, is 55%, when the petrol consumption is 0·29 kg/kWh. Determine (a) the calorific value of the petrol; (b) the corresponding petrol consumption in kg per hour, given that the indicated mean effective pressure is 850 kN/m², and the speed is 2500 rev/min.

Take $\gamma = 1\cdot4$.

(Ans. 41·74 MJ/kg, 19·6 kg/h)
[U.L.C.I.]

26. In an ideal diesel cycle, the volume compression ratio is 12 : 1, and constant pressure combustion is completed at $\frac{1}{20}$ of the stroke. Calculate the pressure and temperature at each point in the cycle, based on air inlet conditions of 17°C and 100 kN/m². Hence determine the thermal efficiency of such a cycle.

What would be the Carnot efficiency for the same temperature range?

Assume $c_p = 1006$ J/kgK, $c_v = 717$ J/kgK.

	kN/m^2		$°C$	
(Ans.	P_1 100	t_1	17	, 0·595, 0·761)
	P_2 3243	t_2	509	
	P_3 3243	t_3	939	
	P_4 185	t_4	261	

[U.L.C.I.]

27. Sketch the pressure-volume diagram for the ideal Carnot (or constant temperature) cycle. Describe the sequence of events in a perfect reciprocating heat engine operating on this cycle and derive an expression for the cycle efficiency.

Determine the Carnot efficiency of (a) a petrol engine in which the lowest and highest temperatures are 80°C and 2060°C respectively; (b) a steam engine supplied with dry saturated steam at 1·4 MN/m² and exhausting to a condenser at 60 kN/m².

(Ans. 84·87%; 23·28%)
[U.L.C.I.]

28. An oil engine working on the ideal Diesel cycle takes in air at 100 kN/m² and 60°C. At the end of the compression stroke the temperature in the cylinder has risen to 690°C. The temperature in the cylinder at the end of combustion is 1400°C. Determine (a) the volume compression ratio of the engine; (b) the maximum pressure in the cycle of operations; (c) the point in the

working stroke at which combustion ceases, and (d) the efficiency of the cycle.

Assume $\gamma = 1\cdot4$.

(Ans. 14·22; 4·113 MN/m²; 5·58%; 60·9%)
[U.L.C.I.]

29. Sketch and describe the pressure-volume diagram for Carnot's cycle when it is applied to an ideal heat engine. Using this diagram, obtain an expression for the efficiency of the cycle.

What are the objections to the use of this cycle as an ideal cycle for a reciprocating internal-combustion engine?

[U.E.I.]

30. Derive an expression, in terms of temperatures only, for the efficiency of the ideal constant volume or 'Otto' cycle.

A petrol engine working on this cycle has a compression ratio of 8·3 : 1. At the beginning of the compression stroke the conditions inside the cylinder are: pressure 90 kN/m², temperature 90°C. The maximum pressure in the cycle is 4 MN/m². Determine the temperature at each of the following points in the cycle: end of compression stroke; beginning of expansion stroke; end of expansion stroke. Use your calculated values of temperature to determine the ideal efficiency of the cycle.

For the working fluid, $\gamma = 1\cdot4$.

(Ans. 573·3°C; 1477°C; 477·6°C; 57%)
[U.E.I.]

31. Sketch the p–V diagram for the ideal air-blast Diesel-engine cycle and obtain an expression for its efficiency in terms of temperatures only.

If the compression ratio of an oil engine, working on the above cycle, is 15 : 1, and air is drawn into the cylinder at 100 kN/m² and 27°C, determine the temperature and pressure of the air at the end of compression. Fuel cut-off occurs at $\frac{1}{12}$ of the working-stroke. Calculate the maximum temperature and the efficiency of the cycle.

Take $c_p = 1006$ J/kgK and $c_v = 717$ J/kgK.

(Ans. 613·1°C; 4·431 MN/m²; 1647°C; 59·5%)
[U.E.I.]

32. Sketch the pressure-volume diagram for the ideal Otto cycle. Derive an expression for the efficiency of this cycle in terms of the volume-compression ratio r and the adiabatic index γ.

The air standard efficiency of a four-stroke petrol engine is based on the expression for the Otto cycle efficiency. Give reasons for this and show that an engine of volumetric

126

compression ratio 9 : 1 has a greater a.s.e. than one of 6 : 1. Take $\gamma = 1 \cdot 4$ for air.

(Ans. 58·5%, 51·2%)
[U.L.C.I.]

33. An oil engine has a compression ratio of 12 : 1, and air is taken into the cylinder at a pressure of 95 kN/m² and at a temperature of 27°C. If the engine may be assumed to be working on the ideal Diesel cycle, determine (a) the maximum pressure in the cycle; (b) the maximum temperature in the cycle if combustion ceases at $\frac{1}{14}$ of the expansion stroke, and (c) the thermal efficiency of the cycle.
Take $\gamma = 1 \cdot 4$.

(Ans. 3·08 MN/m²; 1174°C; 57·84%)
[U.L.C.I.]

34. Define the expressive air standard efficiency and efficiency ratio as applied to an internal-combustion engine.

A four-cylinder, four-stroke petrol-engine of cylinder bore and stroke each equal to 77 mm, has a volume compression ratio of 8·5 : 1. The relative efficiency, or efficiency ratio, referred to the air standard cycle is 58%, when the petrol consumption is 0·24 kg/kWh. Determine (a) the calorific value of the petrol in MJ/kg; (b) the corresponding petrol consumption in kg per hour, given that the indicated m.e.p. is 950 kN/m² when the speed is 3000 rev/min.
Take $\gamma = 1 \cdot 4$.

(Ans. 45 MJ/kg, 8·175 kg)
[U.L.C.I.]

6. Fuels and Combustion

FUELS

Definition

We may define a 'fuel' as a substance composed mainly of carbon and hydrogen, because these elements will combine with oxygen under suitable conditions and give out appreciable quantities of heat. In theory, sulphur should also be counted as a fuel, but the product of burning sulphur is usually an acid which has harmful corroding effects, and these much outweigh the heating value of the substance. The heating or *calorific* values of these fuels are measured by the heat evolved by burning 1 kg of each substance in an adequate supply of oxygen, i.e. in an adequate supply of air, and for the fuels mentioned:

Calorific value of carbon   = 33·7 MJ/kg approximately
Calorific value of hydrogen = 144   MJ/kg   ,,
Calorific value of sulphur   =   9·1 MJ/kg   ,,

Fuels may be broadly classified into solid fuels, liquid fuels and gaseous fuels.

SOLID FUELS

Solid fuels in the first place absorbed their energy from the sun's rays when in the form of vegetation.

Wood consists largely of carbon and hydrogen chemically formed by the action of sunlight, and was at one time used extensively as a fuel. Other solid fuels are derived from partial decomposition of wood and vegetable matter together with the compressing action of superimposed layers as the original wood was buried under earth and rock. The form of the fuel indicates the relative time for which these processes have continued.

Peat. This represents the first stage at which the fuel derived from wood and vegetable matter is recovered from the earth. It is a fibrous watery substance found close to the surface and needs to be dried off before it will burn satisfactorily.

Lignite or Brown Coal represents the next stage, and is an inferior type of coal containing less than 70% carbon and a much higher proportion of moisture than true coal. It has been used extensively on the continent as a low grade fuel. Calorific value about 23 MJ/kg after drying.

Bituminous Coal is the next stage in the development, and is the main fuel mined in Great Britain. It is shiny black in appearance, often showing signs of its vegetable origin, and consists of about 80% carbon and 3% to 5% hydrogen. Calorific value about 31·5 MJ/kg.

Anthracite is regarded as the final stage in the development. It is hard and brittle and consists of about 90% carbon and 3% to 4% hydrogen. It is difficult to ignite but burns with little smoke. Calorific value about 35 MJ/kg.

THE BURNING OF COAL

When coal is heated it breaks up into two distinct fuels:—

(a) Volatile hydrocarbons (i.e. compounds of carbon and hydrogen of the form C_nH_m) which ignite easily and burn with a long yellow flame.

(b) Fixed carbon, which burns with hardly any flame and is difficult to ignite.

Bituminous coal when heated gives off from 20% to 30% volatiles, whereas anthracite contains very few, so that bituminous coal is much easier to ignite than anthracite coal. The distillation temperature of the volatiles, i.e. the temperature at which they leave the coal, is much below the ignition temperature of the fixed carbon, so that it is important to ensure that after distillation they pass through a zone of high temperature and burn, or otherwise they pass away as smoke and soot with a consequent wastage of heat.

Coke is the substance, consisting almost entirely of fixed carbon and ash, which remains after the greater part of the hydrocarbons and moisture has been distilled from the coal as in the production of coal gas. Calorific value about 28 MJ/kg.

THE ANALYSIS OF SOLID FUELS

Analysis of a fuel is carried out to determine its constituents, and two such analyses are employed for solid fuels. In each case, samples must be carefully selected to be truly representative of the bulk, and several tests must be carried out with a number of samples before any conclusions are drawn.

Ultimate Analysis. This analysis is required when important large scale trials are being carried out (e.g. a boiler trial), and is carried out by a chemist. The analysis gives the percentage content by weight of:—

Moisture	Nitrogen	Ash
Oxygen	Sulphur	
Hydrogen	Carbon	

Proximate Analysis. This is a method of obtaining the more important constitutents of a coal and is used, for example, when testing coal purchased in bulk. Each sample of about 2 grammes is carefully weighed into a crucible, and the test is carried out using a small temperature controlled furnace.

(a) Moisture Percentage: The sample is heated for 1 hour at 105°C and then reweighed. Loss of weight gives the moisture content.

(b) Volatile Percentage: The sample is now heated with a loose lid covering, i.e. without access to the air, for about 10 minutes at 955°C and then reweighed. The further loss of weight gives the volatile hydrocarbon content.

(c) Fixed Carbon Percentage: The lid is now removed from the crucible and the sample heated in still air for several hours at 955°C until only ash remains. Reweighing gives the ash, and the further loss of weight is the fixed carbon.

CALORIFIC VALUE OF SOLID FUELS

The calorific value of a fuel can only be satisfactorily determined by experiment, and in the case of solid fuels a 'bomb calorimeter' is used. Whilst a number of different makes of bomb calorimeter are on the market, all of them consist essentially of the same parts, namely:—

1. A strong container called a 'bomb', in which a measured quantity of fuel may be fired in oxygen at high pressure.

2. An electrical arrangement whereby insulated leads are carried to the fuel container inside the bomb, from which a fuse wire dips into the fuel. A switching device is available so that the current may be switched on at a given instant, causing the fuse wire to glow and ignite the fuel.

3. A well-insulated calorimeter containing a known amount of

water into which the bomb may be immersed so as to absorb the heat given out from the fuel. Arrangements must be made for uniform stirring of the water so that the heat is well distributed.

4. A sensitive thermometer which may be well immersed into the water in the calorimeter and give an accurate measurement of the change in water temperature during the experiment.

The procedure is as follows:—

1. Weigh into a crucible about 0·5 grammes of the fuel under test in dry powdered form.

2. Fasten a fuse wire between the conductors and allow the wire to dip into the fuel. (In the case of liquid fuels it is better to keep the wire clear of the liquid, and tie a piece of cotton to the wire and allow the cotton to dip into the fuel. This slows down the rate of combustion and prevents the crucible from being shattered.)

3. Screw on the bomb cap, and fill the bomb slowly with oxygen to a pressure of between 15 and 20 atmospheres.

4. Place the bomb in the calorimeter and surround it with a measured quantity of water up to a level slightly less than the top of the electrical connection.

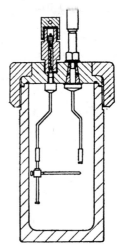

Fig. 57. Section through a Bomb

5. Make the electrical connections ready for firing, and commence stirring. Take temperature readings at intervals of 1 minute.

6. After 10 minutes, when either the water temperature is constant or the change of temperature is regular, fire the bomb and continue stirring, taking temperature readings at short equal intervals until the maximum temperature is reached.

7. Continue stirring, and take temperatures at 1-minute intervals for 10 minutes after the maximum temperature is reached.

The calorific value of the fuel may now be calculated from the general equation:—

Heat released from the fuel = Stored energy gained by water and calorimeter

i.e. mass of fuel × calorific value = (mass of water + water equivalent of calorimeter) × specific heat capacity × rise of temperature of water

131

In order to use this equation, we need to know the water equivalent of the calorimeter, i.e. the mass of water which would absorb heat equivalent to that absorbed by the material of the calorimeter itself. This figure will be given by the makers of the apparatus, but in any case can easily be determined by first carrying out an experiment using a fuel of *known* calorific value, e.g. benzoic acid (29·42 MJ/kg) or Naphthalene (39·72 MJ/kg).

The rise of temperature which is observed will not be the correct value to apply in the formula, for heat may be gained by or lost from the apparatus from the atmosphere during the experiment. It is necessary to make a correction by applying the temperature readings taken during the experiment as shown in Fig. 58.

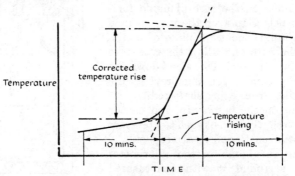

Fig. 58. Water Equivalent of Calorimeter

The calorific value obtained during the experiment is the gross (or higher) calorific value since it includes the latent heat of the steam formed by the combustion of the hydrogen present as it cooled to the temperature of the surrounding water. The net (or lower) calorific value can be calculated, provided the mass of hydrogen in the fuel is known, from the formula.

$$\text{Net c.v.} = \text{Gross c.v.} - (9H_2 \times 2\!\cdot\!453 \text{ MJ})*$$
where H_2 = mass of hydrogen per kg of fuel.

An 'Adiabatic Bomb Calorimeter' (Fig. 59) has been developed by A. Gallenkamp and Co. Ltd., in which the calorimeter is entirely closed in a water jacket in which 'instantaneous' heaters restore equilibrium when the temperature of the jacket falls below that of the calorimeter jacket. In this way, there is claimed to be no heat transfer and it is not necessary to correct the observed temperature rise, hence making the final calculation much simpler.

* Students should refer to B.S. 526/1961 'Definitions of Calorific Value of Fuels' for a new and more complete statement on this subject.

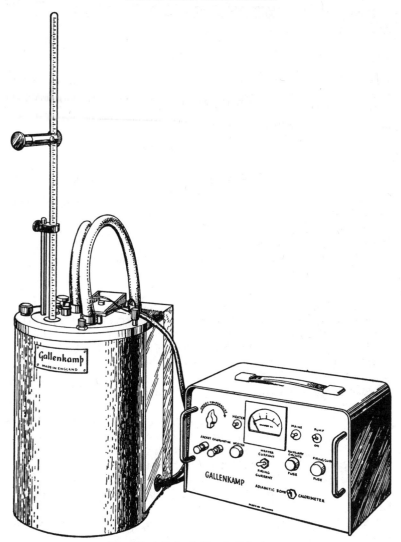

Fig. 59. Adiabatic Bomb Calorimeter

Example 27

'During a test to obtain the calorific value of a sample of coal, the following readings were obtained:—

Mass of fuel burned	$= 0.546$ gramme
Mass of water surrounding the bomb	$= 2800$ grammes
Water equivalent of apparatus	$= 1050$ grammes

133

From plot of temperature readings, corrected temperature rise = 1·056 degC.

Determine the calorific value of the coal.'

Mass of fuel × Calorific value = (Mass of water + Water equivalent of apparatus) × corrected temperature rise × specific heat capacity.

$$0·546 \times \text{Gross c.v.} = (2800 + 1050) \times 1·056 \times 4187$$
$$\therefore \text{Gross c.v.} = 31·12 \text{ MJ/kg}$$

LIQUID FUELS

Most of the liquid fuels used are derived from one or the other of the following sources:—

1. Crude petroleum.
2. By-products from coal gas manufacture.
3. Vegetable matter.

Crude Petroleum. The historical origin of crude petroleum is in some doubt, but it is found in natural oil reserves situated in the earth's crust and from which it is pumped. Crude petroleum is the basis from which the great majority of oil fuels are obtained, the separation of this crude petroleum into the numerous oils used in industry being achieved by distillation (i.e. by heating the crude petroleum and condensing the vapour which evaporates) at various temperatures and pressures.

Below 65°C: All those 'fractions' which distil at temperatures in this range at atmospheric pressure are removed first and are too volatile for use in engine cylinders or heating apparatus.

65°C–220°C: Those fractions which boil between these limits form the *petrols* of industry. Their chemical composition varies widely as does their specific gravity, volatility and ignition temperature.

220°C–345°C: Fractions distilled between these limits form the heating and lighting fuels, i.e. *paraffin* and *kerosene*.

345°C–425°C: The *fuel oils* used in compression-ignition engines and in the furnaces of oil-fired boilers are distilled in this range. The products which now remain if distilled under a vacuum give *mineral lubricating oils*.

Over 425°C: Finally, such products as *greases, paraffin wax, bitumen* or *asphaltum* are obtained at very high boiling temperatures, depending upon the type of crude oil used.

Cracking Distillation is a process whereby the yield of the lighter and more valuable fractions may be increased. The heavier fractions are heated under high pressure in the absence of air, so that the heavy hydrocarbon molecules are decomposed into simple molecules.

By-products from Coal-gas Manufacture. Tar is an important by-product from the manufacture of coal gas, and it may be redistilled to produce valuable fuels like Benzene (C_6H_6) and Toluene (C_7H_8). The important feature of such oils is that they are much less liable to detonation than standard petrols and form a good alternative to petrol in I.C. engines.

Vegetable Matter. Alcohol is formed by fermentation of vegetable matter, and has been widely used on the continent as a commercial fuel. Since it has a high ignition temperature, arrangements have to be made for easy starting when used in I.C. engines, but it has the advantage that detonation is less likely so that higher compression ratios may be used and hence higher thermal efficiencies obtained. An important feature is that supply is practically unlimited, which cannot be said for natural crude petroleum deposits.

PROPERTIES OF LIQUID FUELS

Amongst the properties of liquid fuels, those which concern us are calorific value, flash point, ignition temperature, octane number and cetane number.

Calorific Value of Liquid Fuels. This may be obtained experimentally in the same way as for solid fuels using the bomb calorimeter, except that to avoid shattering the crucible when using liquid fuels the fuse wire should not actually dip into the fuel but should be connected to it by a piece of cotton. In this case, allowance is made for the heating value of the cotton by weighing the piece used, given the calorific value of cellulose = 173 MJ/kg.

Some approximate values of calorific values (gross) for the fuels mentioned are:—

Petrols 46·5 MJ/kg   Benzol 40 MJ/kg
Kerosene 44 MJ/kg
Alcohol 30 MJ/kg
Diesel oil 44 MJ/kg

Flash Points. This is the temperature at which inflammable vapour is given off by a fuel, and is important from storage considerations.

Ignition Temperature. This is the lowest temperature at which

135

the fuel will ignite spontaneously, i.e. without the help of a flame or spark.

Octane Number. Some fuels have a greater tendency to detonation than others, depending upon their ignition temperature and the rate of the combustion reaction. A measure of their tendency to resist detonation is termed the 'octane number' of the fuel. In the determination of the octane number of a fuel, the fuel 'iso-octane' is considered to be 100% anti-detonating and the fuel 'heptane', which makes an engine detonate very readily, is considered to be 100% detonating. The octane number of a fuel is measured by that percentage of octane in an octane-heptane mixture which detonates under the same conditions as the fuel under test.

A petrol engine with an adjustable cylinder head, so that the compression ratio can be altered, and fitted with a 'bouncing-pin' device, which shows the onset of detonation, is used for the test. Running first on the fuel under test, the cylinder head is screwed down and the compression ratio thereby increased until the bouncing-pin device shows detonation to be just beginning. Using this same head setting, the engine is run first on iso-octane and then an increasing percentage of heptane is added to the mixture until detonation occurs again. The percentage of iso-octane in the mixture when detonation re-occurs is termed the octane number of the fuel.

Certain dopes, in particular tetra-ethyl of lead, may be added to a fuel and have been found to increase the resistance to detonation, and thereby improve the octane number of the fuel.

Cetane Number. This is a measure of the suitability of oil fuels for use in compression-ignition engines.

In the operation of high-speed diesels there is always a period of ignition lag between the beginning of fuel injection and the instant when ignition starts, resulting in a severe rise in pressure, and causing the characteristic sound of such engines known as 'diesel knock'. It is desirable that this delay period should be as short as possible to avoid excessive pressure rise. The period of ignition lag can be measured using an accurate indicator, and the fuel to be tested is used under standard conditions in the engine, and the period of ignition lag is measured in degrees of crank angle. A standard fuel consisting of *cetane* which has a very low period of lag and $C_{10}H_7CH_3$ with a large period of lag is then used in the fuel supply line. Starting with 100% cetane, the proportion of $C_{10}H_7CH_3$ is increased until the same period of ignition lag as that of the fuel under test is obtained. The percentage cetane in the standard mixture gives the cetane number for the fuel being tested. Thus, the higher the cetane number, the more suitable is the fuel for compression-ignition work.

136

GASEOUS FUELS

Fuels under this heading include town gas, producer gas and natural gas.

Town Gas. Gas for town supply and on which gas engines are run has, until natural gas from North Sea drillings became available recently, consisted principally of coal gas. This is obtained from the distillation of bituminous coal, and has as its main constituents hydrogen (H_2) and methane (CH_4). Other gases, in particular water gas and blast furnace gas and more recently natural gas, are often mixed with coal gas to form town gas. Water gas is obtained by blowing steam over a bed of burning fuel so that the steam breaks down into its constituents hydrogen and oxygen, the oxygen combining with carbon to form carbon monoxide. Blast furnace gas is a by-product of the smelting process of iron ore. A typical analysis of a town gas is as follows:—

Hydrogen H_2	53·0%	Carbon Dioxide CO_2	2·8%
Methane CH_4	23·0%	Oxygen O_2	0·4%
Carbon Monoxide CO	12·5%	C_mH_n	2·0%
Nitrogen N_2	6·3%		

Gross c.v. of town gas is about 18·5 MJ/m³ at n.t.p.

Producer Gas. This is used, particularly on the continent, in some forms of gas engine, and is formed by partial combustion of anthracite in a generator through which air is drawn by the suction of the engine. A fire is created at the base of the generator, and as the carbon dioxide CO_2, formed by the reaction of the carbon and the oxygen of the air rises through the thick overlying bed of burning anthracite, it is dissociated (i.e. split up) into carbon monoxide CO and oxygen, O_2.

The oxygen liberated combines with more carbon to form more Carbon Monoxide CO. Thus the main constituents of producer gas are carbon monoxide CO, hydrogen H_2, and nitrogen N_2. The calorific value of the gas is low, being of the order of 4·8 MJ (Gross c.v.) per standard cubic metre.

When the air is drawn through the generator by engine suction as described, the plant is called a 'suction producer'. Alternatively, the air can be blown through by a fan or compressor, in which case the plant is a 'pressure producer'.

Natural Gas. This gas is found in porous rocks in oil-bearing regions in many parts of the world and has become a very valuable source of heat for industrial and town use. It has become particularly important in this country since substantial reserves were discovered in the North Sea area, and is scheduled to become the principal

137

supply for domestic and industrial users. The gas can also be compressed and liquefied to make it available for shipment.

Calorific Value of Gaseous Fuels. The calorific value of gaseous fuels may be obtained experimentally using a Boys' gas calorimeter, a diagrammatic arrangement of which is shown in Fig. 60.

The gas to be tested is supplied through a sensitive meter and is pressure-regulated to give a constant supply of gas to the burners at the base of the calorimeter. The hot gases from the burners pass

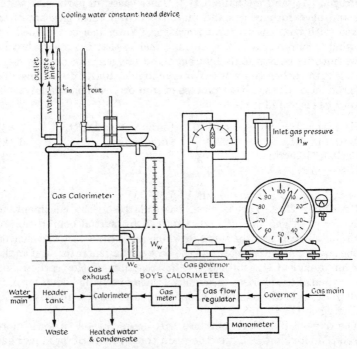

Fig. 60. Boys' Calorimeter

over a set of cooling tubes through which a controlled flow of water is passing, and give up their heat to the cooling water before passing off at the top of the calorimeter. The flow of cooling water is regulated by a constant head weir, and its rise of temperature during its passage through the calorimeter is measured by sensitive thermometers. A small outlet pipe is fitted at the base of the calorimeter from which the condensed steam formed from the combustion of the hydrogen in the gas may be collected.

The calorific value of the gas is given by the general equation:—

Heat released from the gas = Stored energy gained by cooling water

$$\begin{array}{c} \therefore \text{ Volume of gas} \\ \text{burned (m}^3) \\ \text{(corrected to n.t.p.)} \end{array} \times \begin{array}{c} \text{Calorific value} \\ \text{(J/m}^3 \\ \text{at n.t.p.)} \end{array} = \begin{array}{c} \text{Mass of water passing} \\ \text{through cooling tubes} \\ \times \text{rise of temperature} \\ \times \text{specific heat capacity} \end{array}$$

It is essential, when stating the calorific value of a gaseous fuel, to state also the conditions of pressure and temperature to which it refers, for if the pressure and temperature were to be different from those specified, then the volume of gas referred to would also be different. It is usual to state the calorific value of town gas in J per cubic metre at n.t.p. (i.e. 101·3 kN/m² and 15°C). In the test, therefore, it is necessary to measure the temperature and pressure at which the gas is actually supplied, so that the measured volume can be corrected to n.t.p. conditions.

The test is carried out as follows:—

1. Remove the calorimeter from its base and light both burners. Adjust the gas flow so that the meter rotates 1 revolution in 63½ seconds ±5 seconds.
2. Start the water flow in the calorimeter and place it on the base. Adjust the water flow so that 2058 g (±100 g) is collected during 4 meter revolutions.
3. Pour water through the gas effluent holes until it starts to drip from the drain.
4. When conditions are steady:—

   (a) Read inlet water thermometer when meter reads 75.
   (b) At 100, change over the outlet flow from the drain to the measuring vessel, and place condensate collection vessel to drain.
   (c) Read outlet water thermometer when the meter reads 25. Take readings of outlet water at every ¼ turn for 15 readings.
   (d) Read inlet water thermometer at every revolution between 75 and 100.
   (e) After final outlet water temperature reading switch the outlet to drain at 100 on the meter and remove condensate collection vessel. Four revolutions of the meter will have taken place.
5. Other readings to be taken during the test:—

   (a) Gas supply temperature T_1
   (b) Barometer (mm of Hg) h_a
   (c) Inlet gas pressure to meter (mm of H_2O) h_w
   (d) Mass of water in measuring vessel m_w
   (e) Mass of condensate collected during the test m_c

The calorific value of the gas can then be calculated as follows:—

1. Temperature difference of water t_d = Mean outlet temp. – Mean inlet temp.

2. Absolute pressure of gas supplied $p_1 = \left(h_a + \dfrac{h_w}{13 \cdot 6}\right) \times 133 \cdot 3 \text{N/m}^2$

   Volume of gas supplied under test conditions = V_1

   Temperature of gas supplied = T_1 K.
   Volume of gas supplied converted to n.t.p. conditions V_n, can be calculated from

   $$\frac{p_n V_n}{T_n} = \frac{p_1 V_1}{T_1}$$

   where $p_n = 101 \cdot 3 \text{ kN/m}^2$ and $T_n = 288°\text{C}$

3. Heat released by gas = Stored energy gained by water.

   $\therefore V_n \times$ [gross calorific value J/m^3 at n.t.p.] = m_w(kg) $\times t_d$(degC) $\times 4187$

4. Since the condensate (m_c) collected is the condensed steam resulting from the combustion of the hydrogen in gas, the lower calorific value of the gas can be calculated from:—

   Net c.v. = Gross c.v. $- m_c \times 2 \cdot 453 \times 10^6$ J/m^3 at n.t.p.

In the experiment described, no allowance has been made for heat lost to the effluent gases. Full details of corrections are given in *General Notification of the Gas Referees* (H.M. Stationery Office).

Example 28
'In an experiment to determine the calorific value of town gas using a Boys' gas calorimeter, the following readings were obtained:

Mean inlet water temperature	= 20·72°C
Mean outlet water temperature	= 39·36 degC
Mass of water in measuring vessel	= 2130 g
Mass of condensate collected	= 4·93 g
Volume of gas metered	= 10 dm^3
Barometer reading	= 740 mm Hg
Gas supply pressure	= 15 mm water
Gas supply temperature	= 20°C

Determine the values of Gross c.v. and Net c.v. for the gas per cubic metre at n.t.p.'

Temperature difference of water $= 39\cdot36 - 20\cdot72$
$$= 18\cdot64 \text{ degC}$$

Volume of gas converted to n.t.p. conditions:—

$$\frac{p_n V_n}{T_n} = \frac{p_1 V_1}{T_1}$$

$$\therefore V_n = \frac{p_1 V_1}{T_1} \cdot \frac{T_n}{p_n}$$

$$p_1 = \left(740 + \frac{15}{13\cdot6}\right) \text{ mm of Hg}$$

$$= 741\cdot1 \text{ mm of Hg}$$

$$\therefore V_n = \frac{741\cdot1 \times 10 \times 288}{760 \times 293}$$

$$= 9\cdot583 \text{ dm}^3$$

$$= \frac{9\cdot583}{10^3} \text{ m}^3$$

$$\text{moisture collected} = \frac{4\cdot93 \times 100}{1000} \text{ kg}$$

$$= 0\cdot493 \text{ kg}$$

Heat released by gas = Stored energy gained by water.

$$\therefore \frac{9\cdot583}{10^3} \times \text{Gross c.v.} = 2\cdot13 \times 18\cdot64 \times 4187$$

$$\therefore \text{Gross c.v.} = 17\cdot35 \times 10^6 \text{ J/m}^3 \text{ at n.t.p.}$$
$$= 17\cdot35 \text{ MJ/m}^3 \text{ at n.t.p.}$$
$$\text{Net c.v.} = \text{Gross c.v.} - 0\cdot493 \times 2\cdot453$$
$$= 17\cdot35 - 1\cdot209$$
$$= 16\cdot141 \text{ MJ/m}^3 \text{ at n.t.p.}$$

Note: In this experiment the town gas referred to consisted primarily of coal gas. An average calorific value of natural gas from North Sea sources as made available by the Gas Board for domestic use is $37\cdot8$ MJ/m^3 at n.t.p.

COMBUSTION OF FUELS

Anybody can cause fuels to burn—we do it every day when putting a shovelful of coal on the fire—but it takes a good deal of skill and 'know-how' to burn the fuel efficiently, i.e. without waste. There are two very good reasons why every self-respecting engineer should ensure that fuel is burnt efficiently.

1. Fuel is expensive, and inefficient combustion means expensive wastage.
2. Inefficient combustion results in pollution of the atmosphere with noxious gases.

The clue to the correct burning of fuels lies in a knowledge of the combustion equations which show how fuels combine with oxygen to release heat. We shall need to refer to molecular weight of a substance, and so the values are shown again in the following table:—

Element	Symbol	Atomic Weight	Molecular Weight
Carbon	C	12	12
Hydrogen	H_2	1	2
Oxygen	O_2	16	32
Sulphur	S	32	32
Nitrogen	N_2	14	28

When a molecule is made up of atoms of different elements, the molecular weight can be calculated by adding the atomic weights of the elements concerned, for example:—

1 molecule of carbon dioxide (CO_2) consists of 1 atom of carbon + 2 atoms of oxygen.

$$\therefore \text{ Molecular weight of } CO_2 = 12 + (2 \times 16) \quad = 44$$

Similarly   ,,   ,,   ,, carbon monoxide (CO) = 28
,,   ,,   ,, methane (CH_4)   = 16
,,   ,,   ,, sulphur dioxide (SO_2) = 64
,,   ,,   ,, steam (H_2O)   = 18
,,   ,,   ,, petrol (C_7H_{16})   = 100

The burning of carbon to carbon dioxide. This occurs when carbon is burnt with a sufficient supply of oxygen

$$C + O_2 = CO_2$$

This equation states that 1 molecule of carbon combines with 1 molecule of oxygen and produces 1 molecule of carbon dioxide. Inserting the molecular weights, the equation becomes

12 units by mass   +   32 units by mass   =   44 units by mass
of Carbon   of Oxygen   of Carbon Dioxide

i.e. 1 kg of Carbon $+\dfrac{8}{3}$ kg of $O_2 = \dfrac{11}{3}$ kg of CO_2.

Experiment with the bomb calorimeter shows that the calorific value of carbon is about 33·7 MJ/kg, so that

142

1 kg of C needs $\frac{8}{3}$ kg of O_2 and produces $\frac{11}{3}$ kg of CO_2 (and releases 33·7 MJ).

The incomplete burning of carbon to carbon monoxide.

This occurs when carbon is burnt with insufficient oxygen

$$2C + O_2 = 2CO$$

(Note: It is not good enough to write $C + O = CO$, because the smallest quantity of oxygen that can exist alone is a molecule, consisting of two atoms and written O_2.)

Inserting the molecular weights

$$2C + O_2 = 2CO$$
$$(2 \times 12) + 32 = (2 \times 28)$$
$$\text{or } 1 + \frac{4}{3} = \frac{7}{3}$$

i.e. 1 kg of C $+ \frac{4}{3}$ kg of $O_2 = \frac{7}{3}$ kg of CO.

Experiment shows that when 1 kg carbon is burnt to carbon monoxide, only 10·5 MJ of its possible 33·7 MJ is released, the remaining heat being contained in the carbon monoxide which is itself a fuel.

i.e. 1 kg of C taking $\frac{4}{3}$ kg of O_2 produces $\frac{7}{3}$ kg of CO (and releases 10·5 MJ).

The first important lesson should now be clear. Any carbon which is burnt to CO instead of CO_2 is being partly wasted. If CO appears in any chimney or exhaust gas it is direct evidence that fuel is being burnt inefficiently, and means that either insufficient air (i.e. oxygen) is being supplied or that the air which is present is not reaching the fuel because of inadequate mixing.

The burning of carbon monoxide to carbon dioxide.

$$2CO + O_2 = 2CO_2$$

Molecular weights: $(2 \times 28) + 32 = (2 \times 44)$

$$\text{or } 1 + \frac{4}{7} = \frac{11}{7}$$

Thus:—

1 kg of CO needs $\frac{4}{7}$ kg of O_2 and produces $\frac{11}{7}$ kg of CO_2.

To find the heat released by 1 kg of CO, note

1 kg of Carbon burnt to CO_2 releases 33·7 MJ.

143

1 kg of Carbon burnt to $\frac{7}{3}$ kg of CO releases 10·5 MJ

$\therefore \frac{7}{3}$ kg of CO contains a potential $(33 \cdot 7 - 10 \cdot 5)$ MJ

$\therefore$ 1 kg of CO contains a potential $23 \cdot 2 \times \frac{3}{7} = 9 \cdot 94$ MJ

The burning of hydrogen to steam.

$$2H_2 + O_2 = 2H_2O$$
Molecular weights: $(2 \times 2) + 32 = (2 \times 18)$
or $1 + 8 = 9$

i.e. 1 kg of hydrogen needs 8 kg of O_2 and produces 9 kg of steam (and releases 65·41 MJ).

The burning of sulphur to sulphur dioxide.

$$S + O_2 = SO_2$$
Molecular weights: $32 + 32 = 64$
or $1 + 1 = 2$

i.e. 1 kg of S needs 1 kg of O_2 and produces 2 kg of SO_2 (and releases 9·1 MJ).

THEORETICAL AIR SUPPLY

It may be seen from the combustion equations that for complete burning

Carbon requires $\frac{8}{3}$ times its own mass of oxygen

Hydrogen ,, 8 times ,,   ,,   ,,   ,, oxygen
Sulphur requires its own mass of oxygen

Consider the complete combustion of 1 kg of a fuel made of C kg of carbon, H kg of hydrogen, S kg of sulphur, and O kg of oxygen (the remainder being incombustible).

Then oxygen required to burn C kg of Carbon $= \frac{8}{3}C$

,, ,, ,, ,, H kg of Hydrogen $= 8H$
,, ,, ,, ,, S kg of Sulphur $= S$

Total Oxygen required $= \left(\frac{8}{3}C + 8H + S\right)$

The fuel already contains O kg of oxygen which we assume can be used for combustion.

$\therefore$ Theoretical oxygen required for complete combustion of 1 kg of fuel

$$= \left(\frac{8}{3}C + 8H + S - O \right)$$

Now, in most practical cases, the oxygen for combustion of fuels is supplied from the air. Air contains 23% by mass of oxygen, the remaining 77% usually being considered as nitrogen since we neglect the small quantities of other gases present. Hence

Theoretical air supply per kg of fuel burned

$$= \frac{100}{23} \left(\frac{8}{3}C + 8H + S - O \right)$$

With solid fuels it is almost impossible to ensure complete combustion with the theoretical quantity of air owing to the difficulty of effectively mixing them together. It is found necessary to supply more air than is theoretically required, and the quantity of excess air depends on the method of firing the furnace. Too much excess air is also a disadvantage, because it takes heat from the furnace and wastes it to the flue.

THE PRODUCTS OF COMBUSTION

Example 29

'A coal consists of 80% carbon, 4·6% hydrogen, 3% oxygen, and the remainder ash. If 1 kg of the coal is burnt with 20% excess air, determine the mass of the products of combustion and their percentage analysis by mass. Assume air contains 23% oxygen by mass.'

Theoretical air
required per kg of fuel $= \dfrac{100}{23} \left[\dfrac{8}{3}C + 8H + S - O \right]$

$$= \frac{100}{23} \left[\left(\frac{8}{3} \times 0 \cdot 80 \right) + (8 \times 0 \cdot 046) - 0 \cdot 03 \right]$$

$$= 10 \cdot 74 \text{ kg}$$

$\therefore$ Excess air per kg of
fuel $\qquad = 0 \cdot 2 \times 10 \cdot 74 = 2 \cdot 148$ kg
$\therefore$ Actual air supplied/kg $= 12 \cdot 888$ kg

We must now determine the products of combustion as follows:—

(a) CO_2 will have been formed from the combustion of carbon since plenty of air was supplied. From the combustion equation, 1 kg of C produces $\frac{11}{3}$ kg of CO_2.

(b) H_2O will have been formed from the combustion of hydrogen, and the combustion equation states that 1 kg of H_2 produces 9 kg of H_2O.

(c) Since excess air has been supplied, excess oxygen will appear in the products of combustion. The theoretical oxygen will have been used to change C to CO_2, etc., and hence the oxygen in the exhaust products will be given by

$$\text{Excess } O_2 = 0 \cdot 23 \times \text{Excess Air}$$

(d) Nitrogen will also appear in the exhaust products, having gone through the process as an unwelcome passenger—unwelcome because it carries away energy by being heated up from atmospheric temperature to the chimney temperature. Hence, nitrogen in exhaust products is given by $0 \cdot 77 \times$ actual air.

Thus, in the problem: products of combustion per kg of fuel:—

	mass kg	%mass
$CO_2 = \frac{11}{3} \times C = \frac{11}{3} \times 0 \cdot 80 =$	2·933	21·4%
$H_2O = 9H = 9 \times 0 \cdot 046 =$	0·414	3·0%
$O_2 = 0 \cdot 23 \times$ excess air $= 0 \cdot 23 \times 2 \cdot 144 =$	0·493	3·6%
$N_2 = 0 \cdot 77 \times$ actual air $= 0 \cdot 77 \times 12 \cdot 864 =$	9·905	72·0%
Total	13·745	100·0

(percentage analysis is found by expressing each constituent as a percentage of the total mass.

$$\text{e.g. } CO_2 = \frac{2 \cdot 933}{13 \cdot 745} \times \frac{100}{1}\%)$$

We can obtain a check on the total mass of products obtained above, because the same mass of products must come out of the furnace as the mass of fuel and air put into the furnace. (Remember that matter is not destroyed or created.) Hence we can write the general equation for 1 kg of fuel:—

146

1 kg of fuel supplied + mass of air supplied = mass of products of combustion + mass of ash remaining.

In the problem given:—

$1 + 12 \cdot 864 =$ mass of products of combustion $+ 0 \cdot 124$

∴ mass of products $= 13 \cdot 74$ kg, which compares with the value of $13 \cdot 745$ kg obtained previously.

Sometimes the mass and analysis of the *dry* products are required in order to calculate the energy carried away by the exhaust gases. The *dry* products refer to those other than the H_2O (steam), and in this particular problem

Mass of dry products/kg of fuel = 13·327 kg
Mass of H_2O/kg of fuel   = 0·414 kg
————
Total = 13·741 kg
————

ENERGY CARRIED AWAY BY PRODUCTS OF COMBUSTION

We have seen that the products of combustion are made up partly of dry gases and partly of steam. The fuel and air are supplied to the furnace at atmospheric temperature (say 15°C) and leave the chimney at a much higher temperature (160°C to 400°C), and so carry away a substantial amount of energy. The passage of gas through the furnace and chimney is a flow process in which no work is done, hence from the general energy equation the energy lost to the chimney gases is equivalent to the increase in enthalpy of the gases.

$$Q = H_2 - H_1 = mc_p(T_2 - T_1)$$

∴ Energy carried away by dry products per kg of fuel burned

$$= m_g\, c_p(t_{chimney} - t_{atmos.})$$

where m_g = Mass of dry products formed per kg of fuel burned
c_p = Mean specific capacity heat of dry products
$t_{chimney}$ = temperature of exhaust gas at chimney base
$t_{atmos.}$ = temperature of atmosphere in the boiler house

In order to calculate the energy carried away by the steam in the products of combustion, we shall need to know the enthalpy of the

147

steam formed at the temperature of the chimney gases. This will be dealt with in the section on Steam, Chapter 8, page 186.

Example 30

'The fuel supplied to a boiler has the following composition by weight: 85% carbon, 13% hydrogen, 2% oxygen. The air supply is 60% in excess of that theoretically required for complete combustion. If the temperature of the boiler house is 15°C and the temperature of the flue gases is 330°C, calculate the mass of the dry products per kg of fuel burned, and the energy carried away by the dry products per hour when the fuel consumption is 1500 kg per hour. Air contains 23·1% oxygen by mass. Take the specific heat capacity of dry products = 1005 J/kgK.'

$$\text{Theoretical air/kg of fuel} = \frac{100}{23 \cdot 1} \left\{ \frac{8}{3} C + 8H + S - O \right\}$$

$$= \frac{100}{23 \cdot 1} \left\{ \left(\frac{8}{3} \times 0 \cdot 85 \right) \times (8 \times 0 \cdot 13) - 0 \cdot 02 \right\}$$

$$= 14 \cdot 23 \text{ kg}$$

∴ Excess air $= 0 \cdot 6 \times 14 \cdot 23 = 8 \cdot 54$ kg

∴ Actual air $= 22 \cdot 77$ kg

Dry products:—

		Mass
$CO_2 = \frac{11}{3} C = \frac{11}{3} \times 0 \cdot 85 =$		3·117 kg
$O_2 = 0 \cdot 231 \times$ excess air $= 0 \cdot 231 \times 8 \cdot 54 =$		1·97 kg
$N_2 = 0 \cdot 769 \times$ actual air $= 0 \cdot 769 \times 22 \cdot 77 =$		17·5 kg
		22·587 kg

∴ Energy carried away by dry products per hour

$$= m_g c_p (t_2 - t_1)$$
$$= (1500 \times 22 \cdot 587) \times 1005 \times (330 - 15) \text{ J}$$
$$= 10 \cdot 736 \text{ GJ}$$

ANALYSIS OF THE DRY PRODUCTS OF COMBUSTION—CONVERSION OF ANALYSIS BY VOLUME TO ANALYSIS BY WEIGHT

The products of combustion of fuels are gases, and of course it is not a practical proposition to *weigh* the products of combustion. The *Orsat* apparatus, however, provides a convenient method of analysing the dry products by volume. A diagram of this apparatus is shown in Fig. 61. The aspirator bottle E may be used to draw in to the apparatus exactly 100 ml of the exhaust gas under test. The gas is passed through three bottles containing solutions which in turn absorb CO_2, O_2 and CO, the reduction in volume of the gas being measured in each case. After the absorptions have taken place, the gas remaining is assumed to be nitrogen.

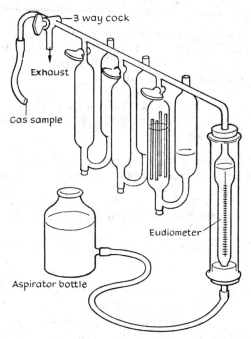

Fig. 61. Orsat Apparatus

We need now to be able to convert the analysis by volume so obtained into an analysis by weight. The procedure is as follows:—

On page 29 we showed the kilogramme-mole to be a quantity of gas having a mass equal to the molecular weight in kilogrammes and occupying the same volume at a given pressure and temperature whatever gas is chosen.

e.g. 1 kmol of oxygen has a mass of 32 kg.

Suppose the Orsat apparatus gave an analysis *by volume* of the dry products from a furnace to be CO_2 8·7%, O_2 5·5%, CO 6·3%, N_2 (remainder) 79·5%. The procedure is shown in the table on page 150.

149

	kmol (a)	Molecular wt (b)	Mass (c) (a) × (b)	% Analysis by Mass (d)
CO_2	0·087	44	3·83	12·9%
O_2	0·055	32	1·76	5·95%
CO	0·063	28	1·76	5·95%
N_2	0·795	28	22·25	75·2%
	1·000 kmol	Mean M for the mixture = 29·60		100·0

Column (a) Shows 1 kmol of gas divided between the constituents in the proportions obtained from the Orsat analysis.

Column (b) Gives molecular weights of constituents, e.g. 1 kmol of CO_2 has a mass of 44 kg.

Column (c) Is the product of columns (a) and (b)
Since 1 kmol of CO_2 has a mass of 44 kg
then 0·087 kmol of CO_2 has a mass of 0·087 × 44 kg
= 3·83 kg.

Note that the total mass of Column (c) is the mass of 1 kmol of the mixture, i.e. the mean molecular weight M for the mixture.

Column (d) The % analysis by mass is now obtained by expressing each constituent as a percentage of the total mass

$$\text{e.g. } \% \text{ weight of } CO_2 = \frac{3·83}{29·6} \times \frac{100}{1} = 12·9\%$$

Example 31

'The exhaust gas from a petrol engine was analysed using an Orsat apparatus with the following results: CO_2 12·5, O_2 3·1, CO 0·5. Convert the analysis to a percentage analysis by weight, and determine the mass of 100 m³ of this gas at 15°C and 100 KN/m².'

Gas	kmol (a)	M (b)	Mass (a) × (b)	Percentage Mass
CO_2	0·125	44	5·50	18·25
O_2	0·031	32	0·99	3·28
CO	0·005	28	0·14	0·47
N_2 (by difference)	0·839	28	23·50	78·00
	1·000	mean M = 30·13 for mixture		100·00

150

Mean molecular weight of gas $= 30 \cdot 13$

$$\therefore \ R = \frac{8314}{30 \cdot 13} = 275 \cdot 9 \ \text{J/kg K}$$

$$\therefore \ m = \frac{pV}{RT}$$

$$= \frac{100 \times 10^3 \times 100}{275 \cdot 9 \times 288}$$

$$= 125 \cdot 8 \ \text{kg}$$

CONVERSION OF ANALYSIS BY WEIGHT TO ANALYSIS BY VOLUME

This is merely the reverse of the procedure just described for the conversion of analysis by volume to analysis by weight and is illustrated in the following example:—

Example 32

'A coal has an analysis by mass of carbon 81%, hydrogen 4·5%, oxygen 3·5% and the remainder ash. Find the theoretical mass of air required for complete combustion of 1 kg of this coal. If the coal is burnt in a furnace with 65% excess air, determine the percentage analysis by mass and by volume of the resulting flue gases. What is the density of the flue gas at s.t.p. (101·3 kN/m² and 0°C)?'

$$\text{Theoretical air/kg of coal} = \frac{100}{23} \left\{ \frac{8}{3} C + 8H + S - O \right\}$$

$$= \frac{100}{23} \left\{ \left(\frac{8}{3} \times 0 \cdot 81 \right) + (8 \times 0 \cdot 045) - 0 \cdot 035 \right\}$$

$$= \underline{10 \cdot 8 \ \text{kg}}$$

$$\therefore \ \text{Excess air} = 0 \cdot 65 \times 10 \cdot 8 = 7 \cdot 02 \ \text{kg}$$

$$\text{Actual air} = 17 \cdot 82 \ \text{kg}$$

Products of combustion:—

			(a) %	(b)	(c) (a)÷(b)	%
		Mass	*Mass*	*M*	*kmol*	*volume*
$CO_2 = \dfrac{11C}{3} = \dfrac{11}{3} \times 0 \cdot 81$	$=$	$2 \cdot 970$	$15 \cdot 88$	44	$0 \cdot 36$	$10 \cdot 69$
$H_2O = 9H_2 = 9 \times 0 \cdot 045$	$=$	$0 \cdot 405$	$2 \cdot 16$	18	$0 \cdot 12$	$3 \cdot 56$
$O_2 \ = 0 \cdot 23$ excess air						
$\quad = 0 \cdot 23 \times 7 \cdot 04$	$=$	$1 \cdot 615$	$8 \cdot 63$	32	$0 \cdot 27$	$8 \cdot 00$
$N_2 \ = 0 \cdot 77 \times$ actual Air						
$\quad = 0 \cdot 77 \times 17 \cdot 82$	$=$	$13 \cdot 721$	$73 \cdot 33$	28	$2 \cdot 62$	$77 \cdot 75$
		$18 \cdot 711$	$100 \cdot 00$		$3 \cdot 37$	$100 \cdot 00$

151

From columns (a) and (c) 100·00 kg of gas occupies 3·37 kmol

∴ mean molecular weight $M = \dfrac{100 \cdot 00}{3 \cdot 37} = 29 \cdot 68$

$$\therefore R = \frac{8314}{29 \cdot 68} = 280 \ \text{J/kgK}$$

From $pV = mRT$

Density $= \dfrac{m}{V} = \dfrac{p}{RT} = \dfrac{101 \cdot 3 \times 10^3}{280 \times 273} = \underline{1 \cdot 324 \ \text{kg/m}^3}$

(Mass per unit volume)

AIR REQUIREMENTS FOR COMBUSTION OF GASEOUS FUELS BY VOLUME AND PRODUCTS OF COMBUSTION

Consider the equation for the combustion of hydrogen into steam:

$$2H_2 + O_2 = 2H_2O$$

This equation states that 2 molecules of hydrogen combine with 1 molecule of oxygen and produce 2 molecules of water. Thus we can interpret the equation in two ways,
either by Mass (inserting the molecular weights)

$$(2 \times 2) + 32 = (2 \times 18)$$

i.e. 1 kg of hydrogen plus 8 kg of oxygen produces 9 kg of steam.
or by Volume

2 kmol of H_2 + 1 kmol of O_2 produce 2 kmol of H_2O

Notice that whilst we can say from the equation by mass that the total mass of the products must equal the mass of fuel and oxygen supplied, the total *volume* of the products is not necessarily the same as the total volume of the gaseous fuel and oxygen supplied. Indeed, in the combustion equation for hydrogen above we see that a total of 3 kmol initially becomes 2 kmol after combustion.

Let us examine now the combustion equations by volume for some gaseous fuels:—

(i) The burning of carbon monoxide to carbon dioxide

$$2CO + O_2 = 2CO_2$$

152

2 kmol of CO + 1 kmol of O_2 produce 2 kmol of CO_2

(ii) The burning of methane (CH_4)

$$CH_4 + 2O_2 = CO_2 + 2H_2O$$

1 kmol of CH_4 + 2 kmol of O_2 produce 1 kmol of CO_2
and 2 kmol of H_2O

(iii) The burning of producer gas

'A producer gas has the following analysis by volume:—

H_2 15%, CH_4 2%, CO 20%, O_2 3%, N_2 54%, CO_2 6%

Assuming that air contains 21% oxygen and 79% nitrogen by volume, determine the theoretical minimum volume of air required to burn 100 m³ of the gas and the corresponding products of combustion measured as percentage by volume.'

The solution is best obtained using a tabular method, and the procedure is as follows:—

(a) The combustion equations by volume give

(i) $2H_2 + O_2 = 2H_2O$
2 kmol + 1 kmol→2 kmol
∴ 15 m³ + 7·5 m³→15 m³

(ii) $CH_4 + 2O_2 = CO_2 + 2H_2O$
1 kmol + 2 kmol→1 kmol + 2 kmol
∴ 2 m³ + 4 m³→2 m³ + 4 m³

(iii) $2CO + O_2 = 2CO_2$
2 kmol + 1 kmol→2 kmol
20 m³ + 10 m³→20 m³

(b) The oxygen (3 m³) already in the gas supplied can be used for the combustion, so that the oxygen required from the air is reduced by this amount.

(c) Notice that air contains 21% oxygen and 79% nitrogen by volume as against 23% oxygen and 77% nitrogen by mass which we have been using previously. (We can write down the equation for air and check these figures by inserting the molecular weights.)

(d) The constituents of the gas N_2 (54 m³) and CO_2 (6 m³) are non-combustibles and cannot be burned further, but they pass unchanged into the products of combustion. The table now gives:—

Gas	kmol m³	kmol of O_2 required	Products of combustion by volume		
			CO_2	H_2O	N_2
H_2	15	7·5		15·0	
CH_4	2	4·0	2·0	4·0	
CO	20	10·0	20·0		
O_2	3	−3·0			
N_2	54	—			54·0
CO_2	6		6·0		
Air	$\dfrac{100}{21} \times 18\cdot5$ $=88\cdot1$	—			$88\cdot1 \times \dfrac{79}{100}$ $=69\cdot6$
Total		18·5	28·0	19·0	123·6

Minimum Air required per 100 m³ of gas = 88·1 m³
Percentage analysis of products by volume

Products	m³	% Volume
CO_2	28·0	16·4
H_2O	19·0	11·1
N_2	123·6	72·5
	170·6	100·0

Notice that 100 m³ of gas required 88·1 m³ of air and the products total 170·6 m³, so that there was a contraction in volume of 17·5 m³

i.e. the *percentage contraction of volume* was $\dfrac{17\cdot5}{188\cdot1} \times \dfrac{100}{1} = 9\cdot3\%$

Example 33
'A gaseous fuel has the following composition by volume:—

48% H_2, 21% CH_4, 1·5% C_6H_6, 19% CO, 6% N_2, 4·5 CO_2

If it is burned with the correct amount of air, determine:—

(a) the percentage contraction of volume resulting from combustion,
(b) the air-gas ratio by volume,
(c) the percentage analysis by volume of the products of combustion,

154

(d) the equivalent molecular weight of the products of combustion.

In addition to the combustion equation for H_2, CH_4 and CO which we have already examined, the combustion equation for C_6H_6 is

$$2C_6H_6 + 15O_2 = 12CO_2 + 6H_2O$$

2 kmol + 15 kmol → 12 kmol + 6 kmol '

Gas	kmol per 100 kmol	kmol of O_2 required	Products of Combustion CO_2	H_2O	N_2
H_2	48·0	24·0		48·0	
CH_4	21·0	42·0	21·0	42·0	
C_6H_6	1·5	11·25	9·0	4·5	
CO	19·0	9·5	19·0		
N_2	6·0	—			6·0
CO_2	4·5	—	4·5		
Air	$\dfrac{100}{21} \times 86\cdot75$ $=413$				$413 \times \dfrac{79}{100}$ $=326\cdot0$
		86·75	53·5	94·5	332·0

Total products $= 53\cdot5 + 94\cdot5 + 332 = 480$ kmol

(a) From the table:—

100 kmol of gas requires 413 kmol of air and produces 480 kmol of products.

∴ Percentage contraction of volume $= \dfrac{513 - 480}{513} \times \dfrac{100}{1}$

$$= 6\cdot4\%$$

(b) Air required per 100 kmol of gas $= 413$ kmol

∴ Air-gas ratio $= 4\cdot13$ to 1

(c) Percentage analysis by volume of the products of combustion

Products	kmol	% kmol
CO_2	53·5	11·14
H_2O	94·5	19·67
N_2	332·0	69·19
	480·0	100·00

(d) Equivalent molecular weight of the products of combustion

Products %	(a) by volume	(b) M	$(a) \times (b)$
CO_2	11·14	44	491
H_2O	19·67	18	355
N_2	69·19	28	1938
	100		2784

i.e. Mass of 100 kmol = 2784 kg

∴ Mass of 1 kmol = 27·84 kg

∴ mean molecular weight, M, for the mixture = 27·84

TO CALCULATE THE GROSS CALORIFIC VALUE OF A FUEL

The only satisfactory way of determining the calorific value of a fuel is the experimental method described using a bomb calorimeter (see page 126). An approximate value can, however, be obtained from a knowledge of the chemical analysis of the fuel, as follows:—

Calculate the gross (or higher) calorific value of an oil fuel having the following composition by mass: carbon 85·5%, hydrogen 12·5%, oxygen 2·0%, taking c.v. of hydrogen = 144 MJ/kg and of carbon = 33·7 MJ/kg.

We first assume that any oxygen present in the fuel is *already* combined with some of the hydrogen, so that the heat release from this part of the hydrogen is not available.

1 kg of H_2 requires 8 kg of O_2 and produces 9 kg H_2O

∴ 0·021 kg of O_2 will have combined with $\dfrac{0·02}{8}$ kg of H_2

$$= 0·0025 \text{ kg}$$

∴ Hydrogen available for heat release = 0·125 − 0·0025

$$= 0·1225 \text{ kg}$$

Calorific value of hydrogen = 144 MJ/kg

∴ Heat release from 0·1225 kg of H_2 = 0·1225 × 144 MJ

$$= 17·64 \text{ MJ}$$

Calorific value of carbon = 33·7 MJ/kg

∴ Heat release from 0·855 kg of C = 0·855 × 33·7 MJ

$$= 28·81 \text{ MJ}$$

∴ Gross c.v. per kg fuel = 17·64 + 28·81

$$= 46·45 \text{ MJ}$$

REVISION EXERCISES—CHAPTER 6

1. In an experiment to determine the water equivalent of a 'Scholes' bomb calorimeter, the following readings were taken:—

$$\text{Weight of naphthalene} = 1 \cdot 012 \text{ g}$$
$$\text{Quantity of water in calorimeter} = 2000 \text{ cm}^3$$

Temperature readings before firing:—

Time (min)	0	1	2	3	4	5	6	7
Temp. °C	20·44	20·44	20·43	20·44	20·44	20·44	20·44 (Fire)	20·43

	8	9	10
	20·43	20·43	20·43

Firing period:—

Time (5 second intervals)	10·05	10·10	10·15	10·20	10·25	10·30	10·35	10·40
Temp. °C	20·44	20·50	20·52	20·80	21·30	21·70	22·10	22·40

10·45	10·50	10·55	11·00	11·05	11·10	11·15	11·20	11·25	11·30
22·85	22·90	22·98	23·05	23·10	23·12	23·42	23·51	23·53	23·55

After firing:—

Time (min)	12·30	13·30	14·30	15·30	16·30	17·30	18·30	19·30
Temp. °C	23·55	23·54	23·53	23·52	23·51	23·50	23·49	23·48

$$\begin{bmatrix} 20 \cdot 30 & 21 \cdot 30 \\ 23 \cdot 47 & 23 \cdot 46 \end{bmatrix}$$

Taking the gross calorific value of naphthalene to be 40 230 kJ, determine the corrected temperature rise and the water equivalent of the apparatus.

(Ans. 3·18°C, 467 g)

2. The percentage composition by weight of a fuel is C 81, H_2 6, O_2 8, remainder ash. If the actual air to fuel ratio by mass used is 17·5 to 1, determine the percentage excess air supplied. Estimate also the heat carried away by the dry products of combustion per kg of fuel if the chimney temperature is 300°C when the boiler house temperature is 15°C. Take c_p for dry products = 1005 J/kgK.

(Ans. 57·2%, 5·13 MJ)

3. Calculate the theoretical mass of air required for complete combustion of 1 kg of fuel containing 88% C and 12% H_2. If the fuel is burnt in a boiler furnace with 50% excess air, calculate the heat carried away per kg of fuel by the dry flue

157

gases, taking the temperature of the air entering the furnace as 18°C and the discharge temperature as 270°C. c_p for dry flue gases = 1005 J/kgK.

(Ans. 14·38 kg, 3·621 MJ)

4. In a boiler trial the fuel analysis was C 88%, H_2 3·6%, O_2 4·8%, S 0·5% by mass, remainder ash. Determine the mass of air required per kg of coal for chemically correct combustion. If the actual air supply is 50% in excess of this, estimate the percentage analysis by mass of the dry flue gases.

(Ans. 11·28 kg, CO_2 18·4%, SO_2 0·06%, O_2 7·40%, N_2 74·10%)
[U.E.I.]

5. The percentage composition by mass of the coal supplied to a boiler is C 81%, H_2 6%, O_2 8% and the remainder ash. Calculate the theoretical mass of air required for complete combustion of 1 kg of coal. If 60% excess air be supplied, determine the total mass of the flue gases produced per kg of coal burnt and state their analysis by mass.

(Ans. 11·12 kg, 18·76 kg, CO_2 15·83%, H_2O 2·88%, N_2 73·05%, O_2 8·21%)

6. An oil fuel consists of 84% carbon and 16% hydrogen by mass. If the air supplied for the combustion of this oil is 20% greater than that needed for complete combustion, estimate the volumetric analysis of the dry exhaust gases.

(Ans. CO_2 11·7%, N_2 84·8%, O_2 3·6%)
[U.E.I.]

7. A fuel of composition C 85%, H_2 4·5%, remainder ash, is burned in a boiler furnace with 50% excess air. Taking $R = 284$ J/kgK for the resulting flue gas, calculate the cross-sectional area of the chimney required if 850 kg of fuel is burned per hour. Take the velocity of the flue gas in the chimney as 3 m/s, chimney pressure and temperature as 96 kN/m² and 340°C respectively.

(Ans. 2·445 m²)

8. An approximate value for the higher calorific value in megajoules of a fuel may be calculated from the formula

$$33 \cdot 7 \, C + 144 \left(H_2 - \frac{O_2}{8} \right) + 9 \cdot 1 \, S$$

Explain the various terms in the formula and calculate the gross c.v. and net c.v. of hexane having the formula C_6H_{14}.

(Ans. 51·68 MJ/kg, 49·23 MJ/kg)

9. The analysis of a sample of dry coal by mass is C 77, H 5, Ash 18. The coal as used contains 10% moisture by mass. Estimate the gross calorific value of the coal as fired, and the percentage

158

by volume of CO_2 in the dry flue gas when the coal is burned with 14 kg of air per kg of coal as used.

$$\text{Take c.v. of carbon} = 33{\cdot}7 \text{ MJ/kg}$$
$$\text{c.v. of hydrogen} = 144 \text{ MJ/kg}$$

(Ans. $29{\cdot}85$ MJ/kg, $12{\cdot}2\%$)

10. A petrol C_7H_{16} is burned with 20% more air than that required for a theoretically correct mixture. Determine, assuming complete combustion,

    (a) the mass of air supplied per kg of fuel.
    (b) the mass of each of the actual exhaust products per kg of fuel burned.
    (c) the percentage analysis by volume of the total products of combustion.

    (Ans. $18{\cdot}36$ kg; CO_2 – $3{\cdot}08$ kg – $10{\cdot}3\%$
    H_2O – $1{\cdot}44$ kg – $11{\cdot}9\%$
    O_2 – $0{\cdot}704$ kg – $3{\cdot}2\%$
    N_2 – $14{\cdot}136$ kg – $74{\cdot}6\%$)

11. A fuel which can be assumed to be CH_4O is burnt with twice the theoretical minimum quantity of dry air required for complete combustion. What proportion of the exhaust gas will be water vapour (a) by weight; (b) by volume?

    (Ans. $8{\cdot}0\%$, $12{\cdot}5\%$)
    [I.Mech.E. Part I]

12. A fuel whose analysis by mass can be taken as 84% carbon and 16% hydrogen is burnt with 10% excess air. What weight and volume of gas at 150 kN/m² and 370°C will pass through the exhaust per kilogramme of fuel burnt?

    (Ans. $17{\cdot}83$ kg, $21{\cdot}1$ m³)
    [I.Mech.E. Part I]

13. Estimate the mass of air to be supplied for the combustion of a fuel containing 75% carbon, 8% hydrogen and 3% oxygen, assuming 50% excess air to be used. What volume of air at 15°C and 100 kN/m² does this represent per pound of fuel?

    (Ans. $17{\cdot}05$ kg, $13{\cdot}6$ m³/kg)
    [I.Mech.E. Part I]

14. Derive an expression for the higher calorific value of a fuel which consists of C% of carbon, H% of hydrogen and O% of oxygen by weight. The appropriate calorific values are:—

    Carbon $33{\cdot}7$ MJ/kg; Hydrogen 144 MJ/kg

    Estimate the higher calorific value of a fuel consisting of 82% carbon, 5% hydrogen and 2% oxygen, the remainder being inert material. What is the minimum quantity of air required for complete combustion of this fuel?

(Ans. 33·55 MJ/kg, 11·16 kg)
[I.Mech.E. Part I]

15. What mass of air will be required per kilogramme of a fuel containing by mass 80% carbon, 5% hydrogen, 3% oxygen and 12% ash, if 50% excess air is to be supplied? What volume of the flue gas will leave the furnace at atmospheric pressure and 370°C per pound of fuel burnt? Take air to contain 23·2% oxygen and 76·8% nitrogen by mass. Universal gas constant = 8314 J/kmolK.

(Ans. 16·17 kg, 30·3 m³ (assuming atmospheric pressure
$= 100 \text{ kN/m}^2$))
[I.Mech.E. Part I]

16. A fuel, with a percentage composition by mass carbon 82, hydrogen 5, oxygen 6, nitrogen 1·0 and remainder ash, is burnt with 60% excess air. If no heat is lost in the process, what is the temperature of the resulting gas mixture (specific heat capacity 1040 J/kgK)? Also determine the composition by mass of the gas mixture. The calorific value of the fuel is 33·7 MJ/kg and the fuel and air are initially at 15°C.

(Ans. 1758°C, CO_2 16·2%
H_2O 2·4%
O_2 8·2%
N_2 73·2%)
[I.Mech.E. Part I]

17. The mass analysis of the fuel as fired during a boiler trial gave the following result:—

Carbon 86%, hydrogen 4%, oxygen 4·6%, sulphur 0·4%, the remainder being ash. If the oxygen in the fuel is in combination with hydrogen in the form of moisture, determine the hydrogen available for combustion.

Calculate the theoretical minimum weight of air required for complete combustion of 1 kg of the fuel, and estimate the dry flue gas analysis by mass, when 50% excess air is supplied. Air contains 23% oxygen by mass.

(Ans. 3·425, 11·25 kg, CO_2 18·05%
O_2 7·41%
N_2 74·5%
SO_2 – trace)
[U.E.I.]

18. The fuel supplied to a boiler contains 78% carbon, 6% hydrogen, 9% oxygen, 7% ash by mass as fired. The air supplied is 50%

in excess of that required for theoretically correct combustion. If the boiler house temperature is 20°C and the flue gas temperature is 315°C, estimate the heat carried away by the dry flue gas per kg of fuel burned. Assume air contains 23% oxygen by mass and the mean specific heat capacity of the dry flue gas is 1005 J/kgK.

Atomic weights: C 12, H 1, O 16

(Ans. 4·88 MJ)

[U.E.I.]

19. The percentage analysis, by mass, of a sample of coal is carbon 80, hydrogen 8, oxygen 5 and the remainder ash. Determine, from the chemical reaction equations, the minimum weight of air required for the complete combustion of 1 kg of this fuel.

If the air supplied is 40% in excess of that required for complete combustion, calculate the percentage analysis, by mass, of the dry products of combustion.

Air contains 23% oxygen by weight.

(Ans. 11·84 kg, CO_2 17·4, O_2 6·5, N_2 76·1)

[U.L.C.I.]

20. Explain what is meant by the higher and lower calorific values of a fuel.

During a test with a bomb calorimeter the following observations were recorded:—

Mass of fuel burned	0·515 g
Mass of water in calorimeter	2420 g
Water equivalent of calorimeter	390 g
Initial temperature of water in calorimeter	13°C
Final temperature of water in calorimeter	14·9°C

If the fuel contains 15% hydrogen (by weight), determine the higher and the lower calorific values of the fuel in British thermal units per pound. The latent heat of the moisture formed by combustion is 2·453 MJ/kg.

What precautions must be taken to obtain an accurate result in an experiment of the above nature?

(Ans. 42·9 MJ/kg, 39·6 MJ/kg)

[U.L.C.I.]

21. A bomb calorimeter was used to determine the calorific value of a sample of coal and the following readings were recorded:—

Mass of coal sample	1·01 g
Mass of water	2500 g
Water equivalent of apparatus	744 g
Temperature rise of water	2·59°C
Temperature correction for cooling	+0·016°C

Make a descriptive line sketch of such a calorimeter and determine the calorific value of the sample in MJ/kg.

Is this the heating value you would expect the fuel to realize when fired to a boiler? Give reasons for your answer.

(Ans. 35·1 MJ/kg)
[U.E.I.]

22. A sample of dry coal has the following percentage analysis by weight: carbon 82, hydrogen 7, oxygen 6, remainder ash. Determine the minimum weight of air necessary for the complete combustion of 1 kg of this fuel.

During a boiler trial using this grade of coal, the air supply was found to be 28% in excess of the minimum necessary for complete combustion. Calculate (a) the actual air supplied per kg of coal; (b) the percentage analysis by mass of the total gaseous products of combustion.

Air contains 23% oxygen by weight.

(Ans. 11·68 kg, 14·95 kg, CO_2 18·9, H_2O 4·0, O_2 4·7, N_2 72·4)
[U.L.C.I.]

23. What do you understand by 'higher calorific value' of a fuel, and how does this differ from the 'lower calorific value'?

The following results were recorded during the determination of the calorific value of coal gas using a gas calorimeter:—

Water collected	2500 g
Inlet temperature	14°C
Outlet temperatuer	28·3°C
Gas consumed	0·01 m³
Gas pressure	89 mm water above atmospheric
Gas temperature	17°C
Barometric pressure	780 mm of mercury

Determine the calorific value of the gas in joules per cubic metre. (A standard cubic metre is measured at 15°C and 760 mm of mercury.)

(Ans. 16·64 MJ/m³)
[U.E.I.]

24. The percentage analysis by mass of a sample of coal is carbon 85, hydrogen 5, oxygen 5, ash 5. Calculate the minimum mass of air required for the complete combustion of 1 kg of this fuel.

If the actual air supplied is 18 kg/kg of coal, calculate the percentage analysis, by weight, of the dry products of combustion.

Air contains 23% oxygen, 77% nigrogen, by weight.

(Ans. 11·35 kg, CO_2 16·85%, N_2 74·87%, O_2 8·28%)
[U.L.C.I.]

25. Define the terms higher calorific value and lower calorific value of a fuel.

A bomb calorimeter was used to determine the higher calorific value of a sample of coal and the following results were recorded:

Weight of coal, 0·915 gf
Weight of water in calorimeter, 2250 gf
Weight of water in bomb, 10 gf
Water equivalent of calorimeter and bomb, 400 gf
Initial temperature of water in calorimeter, 31°C
Final temperature of water in calorimeter, 32·7°C

Calculate the higher and lower calorific values in MJ/kg, assuming the coal has a hydrogen content, by mass, of 16%. At room temperature, latent heat of steam = 2·453 MJ/kg.

(Ans. 20·77 MJ/kg, 17·54 MJ/kg)
[U.L.C.I.]

26. A gaseous fuel has the percentage analysis by volume: hydrogen 16, nitrogen 50, carbon monoxide 28, carbon dioxide 6. Determine the minimum theoretical volume of air required for the complete combustion of 1 m³ of this fuel.

When the fuel is burned, the air supply is at the rate of 1·35 m³ per cubic foot of fuel. Determine the percentage excess air present and the percentage analysis by volume of the dry products of combustion.

Air contains 21% oxygen, by volume.

(Ans. 28·9%; N_2 79·5%, CO_2 17·2%, O_2 3·2%)
[U.L.C.I.]

27. A gaseous fuel has the percentage analysis by volume:

CO 5; H_2 50; CH_4 30; O_2 3; CO_2 2; N_2 10

Determine the minimum volume of air required for the complete combustion of 1 m³ of this fuel.

If the actual air supplied is 5·62 m³/m³ of gas, determine (a) the percentage excess air present, and (b) the percentage analysis by volume of the dry products of combustion.

Air contains 21% O_2, by volume.

(Ans. 39·88%; CO_2 7·0%, O_2 6·4%, N_2 86·6%)
[U.E.I.]

28. A sample of fuel oil has the analysis by mass: carbon 86%, hydrogen 14%. Determine the theoretical higher and lower calorific values of the oil, expressed as MJ/kg.

If the air supplied for combustion of the oil is 24% in excess of the theoretical minimum required for complete combustion,

calculate (a) actual air supplied per kilogramme of oil; (b) the percentage analysis by mass of the total products of combustion.
Air contains 23% oxygen, 77% nitrogen, by mass.
Atomic weights: H 1, C 12, N 14, O 16.

(Ans. Gross c.v. = 49·18 MJ/kg, 18·41 lbf, CO_2 = 16·2%
Net c.v. = 46·09 MJ/kg, H_2O = 6·5%, O_2 = 4·2%,
N_2 = 73·1%)
[U.E.I.]

29. Describe, with the aid of suitable sketches and line-diagrams, a calorimeter which may be used for the accurate determination of the higher and lower calorific values of a gaseous fuel. Outline the method of obtaining these results.

[U.L.C.I.]

30. The percentage analysis, by mass, of a sample of coal is carbon 80, hydrogen 10, oxygen 3, remainder ash. Determine the minimum volume of air at 100 kN/m^2 and 20°C which would be necessary for the complete combustion of 1 kg of this fuel.
If the actual air supply is 24% in excess of the minimum required, determine the percentage analysis by mass of the dry products of combustion.
Air contains 23% of oxygen by mass. For air R = 287 J/kgK.
Atomic weights: H 1, C 12, O 16.

(Ans. 10·68 m^3; CO_2 18·7%, O_2 4·4%, N_2 76·9%)
[U.L.C.I.]

31. A fuel oil has the analysis by mass: carbon, 84%, hydrogen, 16%. Determine the theoretical higher and lower calorific values of this oil.
If the oil is used in a boiler, to which the air supplied for combustion is at the rate of 18·5 kg/kg of oil, determine (a) the percentage excess air present, and (b) the percentage analysis by mass of the dry products of combustion.
Air contains by weight: 23% oxygen, 77% nitrogen. Calorific value of carbon is 33·82 MJ/kg; higher calorific value of hydrogen is 143 MJ/kg.
Atomic weights: H 1, C 12, O 16.

(Ans. 51·28 MJ/kg, 47·76 MJ/kg; 20·9%; CO_2 17%,
O_2 4·1%, N_2 78·9%)
[U.L.C.I.]

32. The volumetric composition of the gaseous fuel supplied to an engine is CH_4 19·5, C_3H_6 1·6, CO 18, H_2 44·4, N_2 13·2, CO_2 3·3. Calculate the volume of air required per cubic metre of fuel for

theoretically correct combustion, the change in volume due to this combustion, and the value of the constant R for the mixture after combustion. Air contains 21% oxygen by volume.

(Ans. 3·686 m³, Percentage reduction 6·48%, 300 J/kgK)
[I.Mech.E.]

7. The Properties of Steam

We have seen that the conversion of heat energy into mechanical work needs a working agent, i.e. a substance which can receive the heat released from a fuel and be used to drive an engine to produce the required work.

Steam is an excellent working agent because of its unique properties:—

(1) It can carry large quantities of stored energy.
(2) It is produced from water which is cheap and readily available.
(3) It can be used for heating purposes after its duty as a working agent is completed.

Steam substance may exist in three states, as a solid, a liquid, or a gas. When used as a working agent in an engine it is most often in the intermediate stage between being a liquid and a gas, i.e. it is a vapour.

A vapour does not obey the laws of ideal gases. Ideal gases are distinguished from real ones by being considered as having molecules which exert no force on each other and occupy no space. In most actual gases, except at very high pressures, these molecular properties are very nearly true and the laws of behaviour of ideal gases may be applied to real gases with reasonable accuracy. In vapours, however, the molecules are heavy and large and consequently their behaviour differs widely from that of ideal gases. The equations for the state of a vapour are so complicated that it is normal to develop tables giving the most common properties.

The energy equations for non-flow and flow processes apply equally to all fluids, and hence may be used for vapours, remembering, however, that the properties of vapours should be read from the tables instead of using the gas laws.

We have seen that the general energy equation is:—

166

Initial stored energy (potential, kinetic, internal) + initial flow work + heat taken in
= Final stored energy (potential, kinetic, internal) + final flow work + external work done.

i.e. expressed in heat units

$$9 \cdot 81 m Z_1 + \frac{m v_1^2}{2} + U_1 + p_1 V_1 + Q = 9 \cdot 81 m Z_2 + \frac{m v_2^2}{2} + U_2 + p_2 V_2 + W$$

For a flow process in which the potential and velocity energies terms may be neglected, the equation reduces to

i.e.
$$(U_1 + p_1 V_1) + Q = (U_2 + p_2 V_2) + W$$
$$H_1 + Q = H_2 + W$$

where H represents enthalpy of the substance.

i.e.
$$Q = W + (H_2 - H_1)$$
Heat taken in or rejected = Work done + change of enthalpy

It follows that for a **flow process** in which no heat is taken in or rejected (e.g. flow through a well-lagged steam turbine)

$$W = H_1 - H_2$$

i.e. Work done by the working agent = $\dfrac{\text{Reduction in enthalpy of the}}{\text{working agent.}}$

For a **non-flow process,** the general equation reduces to

$$U_1 + Q = U_2 + W$$
$$\therefore \ Q = W + (U_2 - U_1)$$

i.e. $\dfrac{\text{Heat taken in}}{\text{or rejected}}$ = Work done + increase of internal energy.

For a non-flow process taking place at constant pressure

$$Q = p(V_2 - V_1) + (U_2 - U_1)$$
$$= (U_2 + p V_2) - (U_1 + p V_1)$$
$$= (H_2 - H_1)$$

THE FORMATION OF STEAM

The properties of steam depend very much on its pressure, so let us first consider the formation of steam at *constant pressure*, being approximately the case when steam is raised in practice in a boiler. Consider 1 kg of water initially at $0°C$ being heated in a vessel fitted with a movable piston such that the pressure in the vessel will remain constant.

During the first stage of heating the temperature of the water *will rise until the water boils at a temperature known as the* SATURATION TEMPERATURE *which depends on the pressure in the vessel.*

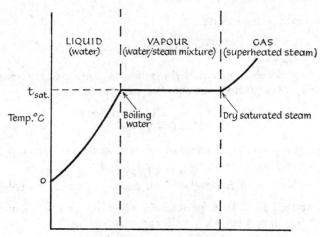

Fig. 62. Formation of Steam

After the boiler temperature is reached, steam begins to be formed, during which time the *temperature remains constant.* Until the point is reached at which all the water is converted into steam, the contents of the vessel will be a mixture of water and steam known as WET STEAM.

Eventually all the water, including those droplets held in suspension, will be evaporated and at this instant, the substance is known as DRY SATURATED STEAM. The substance has, in theory, become a gas at the same temperature as that at which the water boiled.

As heating continues further, the temperature begins to rise again and the steam is now known as SUPERHEATED STEAM, and behaves as a gas. To define the condition of superheated steam it is necessary to state *both* the pressure and temperature of the steam, and the term 'degrees of superheat' is used referring to the amount by which the temperature of the superheated steam exceeds the saturation temperature for the existing pressure.

ENTHALPY OF STEAM

Enthalpy has already been defined as a property of a substance given by the expression

$$H = U + pV$$

For steam, the zero of enthalpy is taken arbitrarily to occur at a temperature of 0°C and tables usually give the values of the enthalpy of 1 kg of boiling water (h_f) and dry saturated steam (h_g) at various pressures.

For 1 kg of water which is not boiling, the specific enthalpy may be estimated from the formula

$$h = 4187t \text{ joules}$$

where t is the water temperature. This neglects the flow work term pV, which is very small except at high pressures. The continuous formation of steam in a boiler is a flow process in which a supply of feed water is pumped into the boiler equivalent to the amount of steam drawn off. Neglecting changes of potential and kinetic energy, the general energy equation states

$$Q = H_2 - H_1$$

i.e. the heat taken in is equal to the change of enthalpy. The heat taken in during the process of converting 1 kg of boiling water into dry saturated steam is known as the *latent heat of evaporation* (l) and is the difference between the specific enthalpies of boiling water and dry saturated steam

i.e. $$l = h_g - h_f$$

In *wet steam*, although all the water has boiled, only part of it has received its latent heat and changed into steam. That fraction of the initial 1 kg of water which has received its latent heat and has become steam, is called the *dryness fraction* (x). The dryness fraction may have any value from zero (corresponding to boiling water) to unity (corresponding to dry saturated steam). The specific enthalpy of wet steam is therefore given by

$$h_{wet} = h_f + x(h_g - h_f)$$

e.g. the enthalpy of 1 kg of steam of dryness fraction 0·9 at 2 MN/m² is

$$h_{wet} = 908 \cdot 6 + 0 \cdot 9 \times 1888 \cdot 6$$
$$= 2608 \cdot 3 \text{ kJ/kg}$$

For *superheated steam*, since, theoretically, the substance has now become a gas, the increase of enthalpy of 1 kg of steam during the superheating stage may be calculated from the equation for the heating of gases at constant pressure, i.e. specific heat capacity × degrees of superheat. Hence the specific enthalpy of superheated steam is given by

$$h_{sup} = h_g + C_p(t_{sup} - t_{sat})$$

169

where C_p = mean specific heat capacity of superheated steam over the range of superheating

t_{sup} = temperature of superheated steam

t_{sat} = temperature of formation

The value of the specific heat capacity of superheated steam is not the same over different ranges of temperature, and the specific enthalpy of superheated steam is best obtained from tables.

E.g. Find the enthalpy of 1 kg of superheated steam at 1 MN/m² and 400°C if the saturation temperature is 179·9°C and the mean specific heat capacity over the temperature range is 2·21 kJ/kg.

$$h_{sup} = 2776\cdot2 + 2\cdot210(400 - 179\cdot9)$$
$$= 2776\cdot2 + 486\cdot4$$
$$= 3262\cdot6 \text{ kJ/kg}$$

VOLUME OF STEAM

The **Specific volume of wet steam** (V_{wet}). 1 kg of wet steam of dryness fraction x consists of x kg of pure steam and $(1 - x)$ kg of water held in suspension,

i.e. $$V_{wet} = xV_g + (1 - x)V_f$$

where V_g is the specific volume of dry saturated steam
and V_f the specific volume of water at formation pressure.

In most cases, except when the steam is very wet, the volume of the water in relation to that of the steam is so small that it may be neglected, and we write

$$V_{wet} = xV_g$$

The **specific volume of superheated steam** (V_{sup}) may be determined direct from tables or calculated from the formula

$$V_{sup} = \frac{230\cdot8(h_{sup} - 1941)}{p} \text{ m}^3/\text{kg}$$

where h_{sup} is the enthalpy in kJ/kg
and p is the pressure in N/m²,
e.g. the volume of 1 kg of steam at 500 kN/m² superheated to 500°C

$$V_{sup} = \frac{230\cdot8(3484 - 1941)}{500 \times 10^3}$$

$$= \underline{0\cdot712 \text{ m}^3/\text{kg}}$$

EXAMPLES ON THE USE OF STEAM TABLES*

The key to success in handling problems is the ability to use the steam tables confidently and accurately. The following examples are worked out so that the student may follow and check them. A similar set of examples are given at the end of the chapter for practice and, as with all skills, 'practice makes perfect'!

Example 34
'Read off the following values:—

(a) Enthalpy per kg of boiling water at 1·0, 6·0, 28, 210, 875 kN/m².
> (Ans. 29, 152, 283, 511, 737·5 kJ/kg)

(b) Enthalpy/kg of dry saturated steam at 100, 155, 300, 490, 900 kN/m².
> (Ans. 2675, 2694·5, 2725, 2748, 2774 kJ/kg)

(c) Volume/kg of dry saturated steam at 14, 450, 750 kN/m².
> (Ans. 10·69, 0·414, 0·2565 m³/kg)

(d) Enthalpy/kg of superheated steam at 50 kN/m² and 100°C, 100 kN/m² and 125°C, 500 kN/m² and 225°C, 2 MN/m² and 400°C, 4 MN/m² and 500°C.
> (Ans. 2683, 2726, 2909, 3248, 3445 kJ/kg)

Example 35
'Calculate the following values:—

(a) Enthalpy/kg of steam at 110 kN/m², 0·9 dry.

$$h_{wet} = h_f + x \cdot h_{fg}$$
$$= 429 + 0·9 \times 2251$$
$$= 2455 \text{ kJ/kg}$$

(b) The change in enthalpy/kg when water at 25°C is converted to steam 0·95 dry at a constant pressure of 1·2 MN/m².

Initial enthalpy of water $= 25 \times 4·187 = 104·7$ kJ/kg
Final enthalpy of wet steam $h_{wet} = h_f + x \cdot h_{fg}$
$$= 798 + 0·95 \times 1986$$
$$= 2684 \text{ kJ/kg}$$

∴ Change in enthalpy $= 2684 - 104·7 = 2579·3$ kJ/kg

---

* The tables used are those in *Thermodynamic and Transport Properties of Fluids* (S I units), compiled by Y. R. Mayhew and G. F. C. Rogers, 2nd edn (Oxford: Basil Blackwell).

(c) The increase in enthalphy when 1 kg of steam initially 0·8 dry is superheated to 350°C at a constant pressure of 2 MN/m².

Initial enthalpy per $kg = h_f + x \cdot h_{fg}$
$$= 909 + 0·8 \times 1890$$
$$= 2421 \text{ kJ/kg}$$

Final enthalpy/kg $= 3138 \text{ kJ/kg}$

∴ Increase in enthalpy/kg $= 3138 - 2421 = \underline{717 \text{ kJ}}$

(d) Find the heat required to raise 2 kg of water initially at 30°C to dry saturated steam at 550 kN/m².

Initial enthalpy/kg of water $= 30 \times 4·187 = 125·7 \text{ kJ}$

At 550 kN/m², $h_g = 2753 \text{ kJ/kg}$

∴ Heat required/kg $= 2753 - 125·7 = 2627·3 \text{ kJ/kg}$

∴ Heat required for 2 kg $= \underline{5255 \text{ kJ}}$

(e) 1 kg of steam is superheated at a pressure of 500 kN/m² to a temperature of 375°C. Determine the mean value of the specific heat during the superheating process.'

At 550 kN/m² temperature of formation $t = 151·8°C$

∴ Degrees of superheat $= 375 - 151·8 = 223·2 \text{ degC}$
$$h_{sup} = 3220 \text{ kJ/kg}$$

but $h_{sup} = h_g + c_p (t_{sup} - t)$
$$3220 = 2749 + c_p (223·2)$$
$$∴ c_p = \underline{2·11 \text{ kJ/kgK}}$$

Example 36

'A boiler is supplied with feedwater at 75°C and produces steam at 2 MN/m² and 275°C. If 70% of the heat of the coal fired of calorific value 30 MJ/kg is given to the steam, calculate the coal consumed in producing 10 000 kg of steam per hour. Find also a suitable diameter for the steam delivery pipe from the boiler assuming a steam speed of 15 m/s.'

Initial enthalpy of feedwater $h = 75 \times 4·187 = 314 \text{ kJ/kg}$

Enthalpy of steam produced $h_{sup} = 2965 \text{ kJ/kg}$

This is a flow process in which steam is formed continuously, and from the general energy equation:—

Heat taken in $Q = H_2 - H_1$ since $W = 0$

∴ Heat given to 1 kg of steam $= 2965 - 314 = 2651 \text{ kJ/kg}$

∴ Heat given to 10 000 kg of steam $= 2651 \times 10 000 \text{ kJ}$

Heat given to steam/kg of coal $= 30 \times 0·7$
$$= 21 \text{ MJ}$$

$$\therefore \text{ Coal consumed} = \frac{2651 \times 10\ 000 \times 10^3}{21 \times 10^6}$$

$$= 1262 \text{ kg/h}$$

Volume kg of delivery steam $v_{\text{sup}} = 0{\cdot}1186 \text{ m}^3$

$$\therefore \text{ Volume of steam flow} = \frac{10\ 000 \times 0{\cdot}1186}{3600} = 0{\cdot}3295 \text{ m}^3/\text{s}$$

Volume of steam flowing through delivery pipe = Area × velocity
$$= A \times 15 \text{ m/s}$$

$$\therefore \quad 0{\cdot}3295 = A \times 15$$

$$A = 0{\cdot}021\ 97 \text{ m}^2$$

$$\frac{\pi D^2}{4} = 0{\cdot}021\ 97$$

$$D = 0{\cdot}167 \text{ m} = 167 \text{ mm}$$

Example 37

'A steam turbine develops 2 MW when admitting steam at 2 MN/m² and 300°C and exhausting at 10 kN/m². Taking the exhaust steam as 0·95 dry, estimate the mass of steam required per hour. Find also the cross-sectional area of the exhaust pipe required. Assume the velocity of the exhaust steam to be 60 m/s.'

From the general energy equation for a flow process

$$Q = W + (H_2 - H_1)$$

neglecting the potential and velocity terms. Hence for no heat taken in or rejected

$$W = (H_1 - H_2)$$
i.e. work done = Change of enthalpy.

Enthalpy of supply steam $h_{\text{sup}} = 3025$ kJ/kg
Enthalpy of exhaust steam $h_{\text{wet}} = h_f + x h_{fg}$
$$= 192 + 0{\cdot}95 \times 2392$$
$$= 2465 \text{ kJ/kg}$$

$$\therefore \text{ Work done/kg of steam} = h_{\text{sup}} - h_{\text{wet}}$$
$$= 3025 - 2465$$
$$= 560 \text{ kJ}$$

Since 2 MW is developed.

$$\text{Work done/s} = 2 \text{ MJ}$$

173

$$\therefore \text{ Weight of steam required/s} = \frac{2 \times 10^6}{0.560 \times 10^6} \text{ kg}$$

$$\therefore \text{ Weight of steam required/hour} = \frac{2}{0.56} \times 3600 = 12\ 857 \text{ kg}$$

For flow of exhaust steam:—

$$\text{Quantity (m}^3\text{/s)} = \text{Area of exhaust pipe (m}^2\text{)} \times \text{velocity (m/s)}$$

$$\text{Volume/kg of exhaust steam} = xv_g$$

$$= 0.95 \times 14.67$$

$$\therefore \text{ Volume of exhaust steam flowing/s} = \frac{2}{0.560} \times 0.95 \times 14.67 \text{ m}^3$$

$$= 49.76 \text{ m}^3$$

$$\therefore \text{ Area of exhaust pipe required} = \frac{49.76}{60} = \underline{0.829 \text{ m}^2}$$

Example 38

'Steam enters a superheater with dryness fraction 0·95 at 4·2 MN/m², and leaves the superheater at 4 MN/m² and 24·7°C superheat. The hot gases flowing over the superheater elements enter at a temperature of 600°C and leave at 480°C. Calculate the required gas flow to superheat 5000 kg of steam per hour. Take c_p for gases = 1005 J/kgK.'

For a flow process $Q = (H_2 - H_1)$ when no work is done.

Enthalpy/kg of steam entering superheater $= 1102 = 0.95 \times 1698$
$$= 2715 \text{ kJ}$$
Enthalpy/kg of steam leaving superheater $= 2882$ kJ
$$\therefore \text{ Heat taken in/kg of steam} = 2882 - 2715$$
$$= 167 \text{ kJ}$$
Heat taken in/h by steam = Heat lost/h by gases
Change of enthalpy/kg of gases $= c_p(t_2 - t_1)$

$\therefore$ Mass of steam/h × heat gained/kg of steam $= m_g c_p(t_2 - t_1)$ where m_g = mass of gas flow/h.

$$\therefore m_g = \frac{5000 \times 167 \times 10^3}{1005 \times (600 - 480)}$$

$$= \underline{6924 \text{ kg/h}}$$

Example 39

'Steam flowing through a long pipeline at a velocity of 30 m/s enters at $1 \cdot 6$ MN/m² dry saturated and leaves at $1 \cdot 5$ MN/m² with dryness fraction $0 \cdot 96$. For a steam flow of 2500 kg/h, calculate the heat loss from the pipeline and select a suitable pipe diameter.'

For a flow process in which no work is done,

$$\text{Heat taken in or rejected } Q = (H_2 - H_1)$$

Enthalpy/kg of steam at entry $= 2794$ kJ

Enthalpy/kg of steam at exit $= 845 + 0 \cdot 96 \times 1947$
$$= 2714 \text{ kJ}$$

∴ Heat loss per lb of steam $= 2794 - 2714 = 80$ kJ

∴ Heat loss/h $= 80 \times 2500 \text{ kJ} = \underline{200 \text{ MJ}}$

Volume/kg of steam at entry $= 0 \cdot 1237$ m³

∴ Volume of steam flow $= \dfrac{2500 \times 0 \cdot 1237}{3600}$ m³/s
$$= 0 \cdot 0858 \text{ m}^3\text{/s}$$

Volume of flow $=$ area $\times$ velocity.

∴ $0 \cdot 0858 = A \times 30$

∴ $A = \dfrac{0 \cdot 0858}{30} = 0 \cdot 002\ 86$ m²

∴ Diameter of pipe $= \sqrt{\dfrac{0 \cdot 002\ 86 \times 4}{\pi}}$
$$= 0 \cdot 060\ 36 \text{ m}$$
$$= \underline{60 \text{ mm}}$$

INTERNAL ENERGY OF STEAM

It is occasionally of value to know the internal energy of steam, and values of the internal energy of 1 kg of boiling water (u_f) and of dry saturated steam (u_g) are given for the range of pressures in columns 4 and 5 respectively of the tables. The specific internal energy of wet steam of dryness fraction x may be calculated from

$$u_{\text{wet}} = u_f + x(u_g - u_f)$$

Alternatively, we have already noted that the specific enthalpy of any substance is given by

$$h = u + pV$$
$$\text{whence } u = h - pV$$

Thus, for 1 kg of:

$$\text{water } u_f = h_f - pV_f$$
$$\text{dry saturated steam } u_g = h_g - pV_g$$
$$\text{wet steam } u_{\text{wet}} = h_{\text{wet}} - pV_{\text{wet}}$$
$$\text{superheated steam } u_{\text{sup}} = h_{\text{sup}} - pV_{\text{sup}}$$

175

pV DIAGRAM FOR WATER-STEAM SUBSTANCE

Fig. 63 is a *pV* state diagram for 1 kg of water-steam substance and may be plotted from the values of pressure and volume given in the tables. This diagram illustrates a number of important points:—

(*a*) For each pressure up to a maximum of 22·12 MN/m² we may plot two values of volume on the chart, the volume of boiling water and the volume of dry saturated steam. The locus of these points form the curve marked in a heavy line with a peak at C. We may

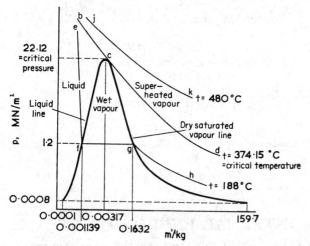

Fig. 63. *pV* State Diagram

observe that the area to the left of the liquid line represents the water state, the area to the right of the dry saturated steam line represents the superheated state, and the area between is the wet vapour state.

(*b*) The curve reaches a maximum at the point C which is called the CRITICAL POINT. Reference to the steam tables shows that at the corresponding critical pressure of 22·12 MN/m², the value of latent heat h_{fg} is zero. At pressures above the critical pressure the liquid and vapour states cannot be distinguished from each other. The corresponding critical temperature is shown from the tables to be 374·15°C, and we may therefore see that steam (or any other gas) cannot be liquified unless its temperature is less than its critical temperature.

(*c*) The curves efgh, bcd, and jk are isothermals (lines of constant

176

temperature) for the temperatures 188°C, 374·15°C and 480°C respectively. The critical temperature for steam is 374·15°C.

Consider the isothermal efgh for the temperature 188°C. The corresponding pressure for this saturation temperature is 1·2 MN/m^2, and the specific volume of boiling water is 0·001 139 m^3/kg. In the liquid state the specific volume reduces only slightly for very large increases in pressure as is shown by the slight slope of the line fe. During the wet vapour phase, however, when latent heat is being taken in and the water is changing into steam, there is a rapid increase in volume whilst the pressure remains constant (fg). The rapid expansion continues if the pressure is reduced during the superheated phase (gh).

Students will find much value in plotting a pV state chart for water-steam substance and a typical isothermal.

EXPANSION OF STEAM—NON-FLOW PROCESSES

As for gases, vapours may be expanded or compressed in a closed system in an infinite number of ways. In the work with which we are concerned such expansions or compressions are usually of the form $pV^n = c$, whence as before

$$\text{Work done} = \frac{p_1 V_1 - p_2 V_2}{(n-1)}$$

Special mention must be made of that expansion such that the index is unity, i.e. $pV = c$. Such an expansion is called a *hyperbolic* expansion. Note that it is *not* at constant temperature (i.e. isothermal) as with gases. An isothermal process occurs for wet steam when the pressure is constant, i.e. during the addition or rejection of latent heat.

$$\text{Work done during a hyperbolic process } W = pV \log_e \frac{V_2}{V_1}$$

From the energy equation for a non-flow process, the heat flow is given by

$$Q = W + (U_2 - U_1)$$

Example 41

'One kilogramme of steam is compressed in a closed system according to the law $pV^{1·2} = c$ from a pressure of 100 kN/m^2 and dryness fraction 0·80 to a final pressure of 220 kN/m^2. Find

(a) the final condition of the steam;
(b) the external work done on the steam expressed in heat units;
(c) the heat flow across the boundaries of the system.'

177

(a) $p_1v_1{}^{1\cdot2} = p_2v_2{}^{1\cdot2}$
   $\therefore \; p_1(x_1v_{g1})^{1\cdot2} = p_2(x_2v_{g2})^{1\cdot2}$

$$\therefore \; x_2 = \left(\frac{p_1}{p_2}\right)^{\frac{1}{1\cdot2}} \cdot \frac{x_1v_{g1}}{v_{g2}}$$

$$= \left(\frac{100}{220}\right)^{\frac{1}{1\cdot2}} \times \frac{0\cdot8 \times 1\cdot694}{0\cdot810}$$

$$= \underline{0\cdot8674}$$

(b)   Work done $= \dfrac{p_1v_1 - p_2v_2}{(n-1)}$

$$= \frac{10^3}{0\cdot2} \times 100 \times 0\cdot8 \times 1\cdot694 - 220 \times 0\cdot8674 \times 0\cdot810$$

$$= -95\,000 \text{ J} = \underline{-95 \text{ kJ}}$$

(Negative since work is done *on* the steam and hence flows *into* the system.)

(c) This is a non-flow process and hence

$Q = W + (U_2 - U_1)$
$u_1 = h_1 - p_1v_1$
   $= (h_{f1} + x_1h_{fg1}) - p_1(x_1v_{g1})$
   $= 417 + 0\cdot8 \times 2258 - 0\cdot1 \times 10^3 \times 1\cdot694 \times 0\cdot8$
   $= 2088 \text{ kJ}$
$u_2 = (h_{f2} + x_2h_{fg2}) - p_2x_2v_{g2}$
   $= 518 + 0\cdot8674 \times 2193 - 0\cdot22 \times 10^3 \times 0\cdot81 \times 0\cdot8674$
   $= 2265 \text{ kJ}$
$\therefore \; Q = W + (u_2 - u_1)$ for 1 kg of steam
   $= -95 + (2265 - 2088)$
   $= 82 \text{ kJ/kg}$

i.e. heat is taken in through the containing walls.

Example 42
'2 kg of dry saturated steam at an initial pressure of 700 kN/m² expands hyperbolically in a closed system to a final pressure of 100 kN/m². Determine the final state of the steam and the heat flow across the cylinder walls, stating in which direction it occurs.'

(a) For hyperbolic expansion $p_1v_1 = p_2v_2$

$$\therefore \ p_1v_{g1} = p_2v_{sup2}$$

$$\therefore \ v_{sup2} = \frac{700}{100} \times 0 \cdot 273 = 1 \cdot 911 \ \text{m}^3/\text{kg}$$

$$\text{Now} \ v_{sup} = \frac{230 \cdot 8(h_{sup} - 1941)}{p}$$

$$\therefore \ h_{sup2} = \frac{10^5 \times 1 \cdot 911}{230 \cdot 8} + 1941$$

$$= 2768 \ \text{kJ}/\text{kg}$$

Hence from steam tables, the temperature of the superheated steam at 100 kN/m² is 146°C approximately.

(b) For a non-flow process $Q = W + (U_2 - U_1)$ and for 1 kg of steam:—

from tables of Internal Energy,

$$u_1 = 2571 \cdot 1 \ \text{kJ}/\text{kg}$$

$$u_2 = 2506 + \frac{46}{50}(2583 - 2506)$$

$$= 2506 + 71$$

$$= 2577 \ \text{kJ}/\text{kg}$$

$$W = p_1v_1 \log_e \frac{v_2}{v_1}$$

$$= 700 \times 10^3 \times 0 \cdot 273 \log_e 7 \ (\text{J})$$
$$= 371 \cdot 8 \ \text{kJ}/\text{kg}$$
$$Q = W + (u_2 - u_1)$$
$$= 371 \cdot 8 + (2577 - 2573)$$
$$= 375 \cdot 8 \ \text{kJ}/\text{kg}$$

$\therefore$ For 2 kg of steam, $Q = 751 \cdot 6$ kJ, and heat is taken in by the steam.

THROTTLING

When steam is forced through a small orifice under pressure, the steam is said to be 'throttled'. For example, throttling occurs when steam passes through a partially open valve. At the orifice, eddies are formed which eventually reconvert their energy into less useful energy at the lower pressure, so that *throttling is always a wasteful operation* which lowers the quality of the energy.

179

Fig. 64 shows a fluid flowing along a pipe from a higher pressure p_1 to lower pressure p_2 and passing through a throttling orifice.

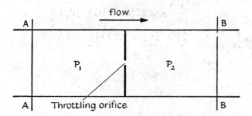

Fig. 64. Throttling

Consider the energy conditions at AA and BB.
For a flow process:—

$$9{\cdot}81\ mZ_1 + \frac{mv_1{}^2}{2} + U_1 + p_1V_1 + Q = 9{\cdot}8\ mZ_2 + \frac{mv_2{}^2}{2} + U_2 + p_2V_2 + W$$

There is no change of potential energy, no external work is done, and if the pipe is well lagged there is no heat taken in or rejected. Furthermore, in practical cases the velocities at AA and BB are small and little different, hence we may write

$$U_1 + p_1V_1 = U_2 + p_2V_2$$
$$\text{i.e. } H_1 = H_2$$

that is to say, the *enthalpy remains constant*.

The steam tables show that the enthalpy of dry saturated steam h_g increases with pressure ($3{\cdot}2$ MN/m²). Suppose 1 kg of wet steam at a high pressure is throttled to some lower pressure. Since the enthalpy/kg of the steam at the low pressure after the throttle is to be the same as the enthalpy/kg at the high pressure before the throttle, the steam must be dried out to some degree.

Example 43

'1 kg of steam initially 0·8 dry is throttled from a pressure of 600 kN/m² to a pressure of 100 kN/m². What will be the dryness fraction of the steam after throttling?'

The steam tables give:—

p kN/m²	h_f kJ/kg	h_{fg} kJ/kg
600	670	2087
100	417	2258

Enthalpy/kg after throttle = enthalpy/kg before throttle. If steam is still wet after throttle:—

$$(h_{f2} + x_2 h_{fg2}) = (h_{f1} + x_1 h_{fg1})$$
$$417 + (x \times 2258) = 670 + (0.8 \times 2087)$$
$$\therefore x_2 = 0.85$$

This property whereby wet steam can be dried and sometimes superheated by throttling is used in an instrument for measuring the dryness fraction of steam.

THE THROTTLING CALORIMETER

Sample steam is taken from a pipeline as shown, and passed through a throttling orifice to the atmosphere. Measurements are made of the pressure before the throttle p_1, and the pressure and temperature after the throttle p_2 and t_2. *Provided the steam after*

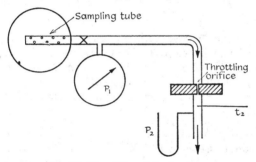

Fig. 65. Throttling Calorimeter

throttling is superheated, its enthalpy may be read from the steam tables knowing the final temperature and pressure p_2. The values n_{f1} and h_{fg1} can be read for the corresponding pressure before the throttle p_1.

Thus the unknown dryness fraction of the supply steam x_1 can be calculated from:—

Enthalpy/kg after throttle = enthalpy/kg before throttle.

$$\therefore h_{sup2} = h_{f1} + x_1 h_{fg1}$$
$$\therefore x_1 = \frac{h_{sup2} - h_{f1}}{h_{fg1}}$$

Example 44

'Steam in a pipeline at 900 kN/m² gauge is sampled by a throttling calorimeter which gives a temperature reading of 110°C after the throttle when the atmospheric pressure is 100 kN/m².

Find the dryness fraction of the sample steam.'

From steam tables,

$$h_{sup2} = 2696 \text{ kJ/kg}$$

Before the throttle, 900 kN/m^2 gauge $= 1 \text{ MN/m}^2$

At 1 MN/m^2, $h_{f1} = 763 \text{ kJ/kg}$

$$h_{fg1} = 2015 \text{ kJ/kg}$$
$$h_{sup2} = h_{f1} + x_1 h_{fg1}$$
$$2696 = 763 + (x_1 \times 2015)$$
$$\underline{x_1 = 0.96}$$

This method will not give results unless the final steam is superheated, for if the final steam is wet its enthalpy cannot be determined from the steam tables. The limiting condition is when the final steam is just dry saturated, and a throttling calorimeter alone will only give results for sample steam dryness fraction greater than about 0.95.

Example 45

'The pressure of the steam after passing the throttling orifice of a throttling calorimeter is 100 kN/m^2. If the initial steam supply pressure was 1.05 MN/m^2, what would be the least dryness fraction of the steam which could be measured?'

The final steam must be just on the point of superheating, i.e. just dry saturated.

Enthalpy of dry sat. steam at $100 \text{ kN/m}^2 = 2675 \text{ kJ/kg}$

For supply steam at 1.05 MN/m^2, $h_{f1} = 772.0 \text{ kJ/kg}$

$$h_{fg1} = 2007.5 \text{ kJ/kg}$$
$$h_{g2} = h_{f1} + x_1 h_{fg1}$$
$$2675 = 772.0 + (x_1 \times 2007.5)$$
$$\underline{x_1 = 0.949}$$

THE COMBINED SEPARATING AND THROTTLING CALORIMETER

In order to deal with steam of *any* dryness fraction, a separator is placed before the throttling orifice. By redirecting the steam, the separator causes it to throw off most of its moisture before throttling so as to ensure that it becomes superheated after throttling. The steam emerging from the throttle is condensed in a small surface condenser.

Steam is allowed to pass through the apparatus for a warming-up period in order to obtain steady conditions, and before measurements are taken the surplus moisture is drained off the separator. During a test of say 20 minutes, measurements are made of the mass of steam passing through the throttling orifice $(\dot{m}_s)$ and the corresponding mass of water collected in the separator $(\dot{m})$, together with mean readings of supply pressure (p_1), pressure after the throttle (p_2) and temperature after the throttle (t_2).

After the throttling orifice $h_{sup2} = h_{f1} + x_1 h_{fg1}$
  where $x_1 =$ dryness fraction of steam after the separator

Since $\dot{m}_s$ kg of steam passes through the throttle, at section BB before the throttle there is $\dot{m}_s$ kg of wet steam of which $x_1 \dot{m}_s$ is pure steam and $(1 - x_1)\dot{m}_s$ is water. At the separator, $\dot{m}$ kg of moisture is collected.

∴ At section AA before the separator there is $(\dot{m}_s + \dot{m})$ kg of wet steam of which $x_1 \dot{m}_s$ is pure steam and $[(1 - x_1) \dot{m}_s + \dot{m}]$ is water.

$$\therefore \textbf{ Initial dryness fraction } x = \frac{x_1 \dot{m}_s}{(\dot{m}_s + \dot{m})}$$

Example 46

'The steam condensed after passing through the throttling orifice of a combined separating and throttling calorimeter weighs 25 kg, and during the same time 3·4 kg of water is drained from the separator. Other measurements taken are initial steam pressure = 0·65 MN/m², final steam pressure = 0·10 MN/m², final steam temperature 125°C. Estimate the initial dryness fraction of the steam.'

$$\therefore h_{sup2} = 2726 \text{ kJ/kg}$$
$$\text{After the throttle } h_{sup2} = h_{f1} + x_1 h_{fg1}$$
$$\therefore 2726 = 683 \cdot 5 + (x_1 \times 2077)$$
$$x_1 = 0 \cdot 9833$$
$$\text{Initial dryness fraction } x = x_1 \frac{m_s}{m_s + m}$$
$$= \frac{0 \cdot 9833 \times 25}{28 \cdot 4}$$
$$= \underline{0 \cdot 857}$$

The enthalpy of super-heated steam may be calculated from

$$h_{sup} = h_g + c_p(t_{sup} - t)$$

using a mean value of specific heat $c_p = 0 \cdot 5$.

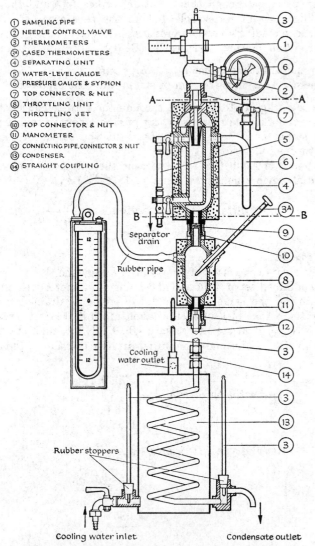

① SAMPLING PIPE
② NEEDLE CONTROL VALVE
③ THERMOMETERS
③A CASED THERMOMETERS
④ SEPARATING UNIT
⑤ WATER-LEVEL GAUGE
⑥ PRESSURE GAUGE & SYPHON
⑦ TOP CONNECTOR & NUT
⑧ THROTTLING UNIT
⑨ THROTTLING JET
⑩ TOP CONNECTOR & NUT
⑪ MANOMETER
⑫ CONNECTING PIPE, CONNECTOR & NUT
⑬ CONDENSER
⑭ STRAIGHT COUPLING

Separator drain

Rubber pipe

Cooling water outlet

Rubber stoppers

Cooling water inlet

Condensate outlet

Fig. 66. Combined Separating and Throttling Calorimeter
(W. Sisson & Co. Ltd.)

184

REVISION EXERCISES—CHAPTER 7

1. Use the steam tables to find the following:—

   (a) Enthalpy and volume per kg of steam of dryness fraction 0·85 at a pressure of 0·7 MN/m².

   (b) Enthalpy per kg of steam at a pressure of 2 MN/m² and temperature 400°C.

   (c) Dryness fraction of steam at 0·32 MN/m², having enthalpy equal to 2600 kJ/kg.

   (d) The heat required to convert 2 kg of water from 40°C to dry saturated steam at a pressure of 0·6 MN/m².

   (Ans. (a) 2454 kJ, (b) 3248 kJ, (c) 0·941, (d) 5179 kJ)

2. Use the steam tables to calculate the following:—

   (a) The internal energy of 1 kg of steam of dryness fraction 0·9 at a pressure of 0·35 MN/m².

   (b) The internal energy of 1 kg of steam at a pressure of 0·5 MN/m² and temperature 300°C.

   (c) The temperature, specific volume and value of the mean specific heat capacity over the superheating range of superheated steam at a pressure of 2·0 MN/m² which has an enthalpy of 2965 kJ/kg.

   (Ans. (a) 2253 kJ, (b) 2804 kJ, (c) 275°C, 0·1185 m³, 2·643 kJ/kgK)

3. Steam enters a turbine at a pressure of 4 MN/m² and temperature 400°C and exhausts at 10 kN/m² and dryness fraction 0·9. Estimate the power developed for a steam flow of 3000 kg/h and the cross-sectional area of the exhaust duct required for an exhaust steam velocity of 50 m/s.

   (Ans. 724·3 kW, 0·22 m²)

4. A boiler is required to produce 3000 kg of dry saturated steam per hour at 1 MN/m² from feedwater at 65°C. If 70% of the heat of the fuel fired is given to the steam, find the mass of the fuel required per hour. Take calorific value of fuel as 32 MJ/kg. Find also a suitable pipe diameter to draw off this steam from the boiler, so that the steam velocity shall not exceed 25 m/s.

   (Ans. 335·6 kg, 100 mm pipe (90·8 mm diam.))

5. A boiler is required to produce steam at 1 MN/m² and superheated to 300°C. The feedwater used is at 50°C and the steam enters the superheated tubes at a dryness fraction of 0·92. Estimate the proportion of the heat supplied taken in in the superheater.

   (Ans. 15·3%)

6. A throttling calorimeter is used to sample the steam supplied from a boiler to a superheater at a pressure of 1 MN/m². The throttling calorimeter discharges to an atmospheric pressure of 0·1 MN/m² and the temperature of the steam after the throttle is 125°C. Estimate the dryness fraction of the steam entering the superheater.

The steam leaves the superheater tubes at a pressure of 1 MN/m² and a temperature of 275°C. If the temperature drop of the gases flowing over the superheater tubes is 40°C, find the mass of gas flow per kg of steam produced taking c_p for gases = 1007 J/kg deg C.

(Ans. 0·974, 6·753 kg)

7. A low-pressure evaporator is supplied with dry steam at 0·2 MN/m² and the condensate is removed from the coils at 60°C. The feedwater enters the evaporator at 30°C. The pressure in the evaporator is 0·1 MN/m² and the steam formed leaves 0·9 dry. If 30 kg of water are evaporated per m² of heating surface per hour, calculate the heating surface of the coils and the consumption of heating steam when evaporating 110 kg/h.

(Ans. 3·667 m², 104 kg)

[U.E.I.]

8. The steam leaving the superheater of a boiler is at a pressure of 2 MN/m² and at a temperature of 350°C. The wet steam entering the superheater at 2 MN/m² is sampled by means of a throttling calorimeter and the temperature after throttling is 125°C. The pressure in the calorimeter is 0·1 MN/m². Calculate, per kg of steam passing through the superheater:—

(a) the heat received;
(b) the increase in volume.

(Ans. 412 kJ, 0·0428 m³)

[L.U.]

9. The total mass of steam present in a cylinder is 0·01 kg; the measured pressures and volumes at two points A and B are:—

Point	Pressure MN/m²	Volume m³
A	0·5	0·0020
B	0·1	0·0055

For these two points determine

(i) the dryness of the steam;
(ii) the internal energy of the steam.

(Ans. 0·533; 0·325)

[L.U.]

10. Explain how a combined separating and throttling calorimeter may be used to estimate the dryness of steam in a steam main. Make suitable sketches to show the construction of the apparatus and how the sample is taken.

In such an arrangement, the steam main pressure is 0·7 MN/m^2 and the temperature after throttling is 120°C. The pressure in the throttling calorimeter is 0·1 MN/m^2. At the separator 0·11 kg of water is trapped and 2·14 kg of steam are passed through the throttling calorimeter. Determine the dryness of the steam in the steam main using the steam tables to find the enthalpy per kg of the superheated steam.

(Ans. 0·929)

[L.U.]

11. What is the final condition of 1 kg of steam at the end of each of the following operations, if the initial condition of the steam in each case is 0·95 dry at a pressure of 1 MN/m^2?

  (a) The temperature is reduced at constant volume to 91·8°C.
  (b) The steam loses 60 kJ at constant pressure.
  (c) The steam does work equivalent to 120 kJ in an engine and leaves at a pressure of 0·1 MN/m^2.
  (d) The steam is throttled until it is just dry and saturated.
  (e) The steam receives 100 kJ at constant pressure.

(Ans. (a) $x = 0·083$; (b) $x = 0·920$; (c) $x = 0·948$;
(d) pressure = 0·1 MN/m^2; (e) dry. sat.)

12. At a point just after cut-off in a steam engine cylinder, the steam pressure was 1·1 MN/m^2 and the dryness 0·85. After expansion, at a point just before release, the pressure was 0·28 MN/m^2 and the dryness 0·80. Assuming that the expansion obeys a law $pV^n = c$, determine the heat flow per kg across the cylinder walls, and state its direction.

(Ans. 0·2 kJ (added))

13. A quantity of steam having a dryness fraction of 0·9 and occupying a volume of 2 m^3 at a pressure of 0·2 MN/m^2 is compressed in a closed system according to the law $pV^n = c$ until the pressure and temperature are respectively 1 MN/m^2 and 200°C. Determine the value of the compression index n, the work done, and the heat received or rejected by the steam during the compression. State the direction of heat transfer.

(Ans. $n = 1·19$, $Q = 273·4$ kJ (rejected))

14. Define 'Enthalpy' of steam and show that the value of this quantity remains unaltered when the steam passes through a throttling process.

187

Steam at 0·6 MN/m² in passing through a throttling process has its pressure reduced to 0·1 MN/m² and is then dry and saturated. Calculate the initial dryness of the steam and the change of internal energy during the process.

(Ans. 0·96, – 13 kJ/kg)

[L.U.]

15. The following observations were made whilst a combined separating and throttling calorimeter was used to find the dryness fraction of steam supplied at 0·55 MN/m². Temperature after throttling 110°C; pressure after throttling 0·1 MN/m²; mass of water collected in the separator 0·14 kg; mass of steam passed through the throttle 9·2 kg; Calculate the initial dryness of the steam.

(Ans. 0·958)

16. (a) Describe with the aid of simple diagrams the operation of the throttling calorimeter.

(b) An engine is supplied with steam from a boiler through a long pipeline. The steam leaving the boiler is at 1 MN/m² and 220°C. At the engine end of the pipeline some of the steam is passed through a throttling calorimeter. The steam enters the calorimeter at a pressure of 1 MN/m² and is throttled to 0·1 MN/m² and temperature 135°C. Determine

(i) The dryness fraction of the steam at the engine.

(ii) The heat lost per kg of steam during passage through the pipe.

(Ans. 0·984, 129 kJ/kg)

17. Feedwater at a temperature of 50°C enters an economizer and leaves it at 127°C. It then passes into a boiler which generates steam at a pressure of 1·5 MN/m² and dryness fraction 0·98. The steam on leaving the boiler enters a superheater and on leaving the superheater the temperature of the steam is 300°C. The fuel used per hour is 725 kg and the steam produced is 5100 kg/h. Calorific value of the fuel used is 26·5 MJ/kg.

Determine the heat utilized per pound of fuel (a) in the economizer; (b) in the boiler; (c) in the superheater, and find the overall efficiency of the steam-raising plant.

(Ans. 2269 kJ; 15 630 kJ; 2012 kJ, 75·11%)

[U.L.C.I.]

18. Wet steam at a pressure of 1·2 MN/m² flows along a 150 mm diameter pipe at the rate of 4200 kg/h. The velocity of the steam in the pipe is 9 m/s. Determine the dryness fraction of the steam.

The steam passes through a separator where 300 kg/h of water are collected, and then through a valve where the steam is throttled to 0·24 MN/m². Determine the condition of the steam after throttling.

(Ans. 0·835; 0·940)
[U.L.C.I.]

19. Make a neat sketch of a combined throttling and separating calorimeter, and explain how such an instrument can be used to find the dryness fraction of the steam leaving a boiler.

In such a test the following results were obtained:—

Boiler pressure 1·25 MN/m². Pressure of steam after throttling 0·1 MN/m², temperature of steam after throttling 105°C, water collected from separator 0·16 kg per minute, discharge from the throttle calorimeter 7 kg per minute. Calculate the dryness fraction of the steam leaving the boiler. Take the specific heat capacity of the superheated steam as 2·1 kJ/kg deg C.

(Ans. 0·929)
[U.E.I.]

20. A cylinder contains 0·2 kg of steam at 1·3 MN/m² and 0·9 dry. If the steam is expanded until the pressure falls to 0·6 MN/m² the law of expansion being pV = constant, determine,

(a) the temperature at the end of the expansion;

(b) the heat flow into or out of the steam during the expansion, stating the direction of flow.

(Ans. 158·8°C, 7·29 kJ taken in)
[U.E.I.]

21. Describe how you would determine the dryness fraction of a sample of steam using a combined separating and throttling calorimeter.

From such a test it was found that the pressure in the steam main was 0·75 MN/m² and the pressure and temperature after throttling were 0·1 MN/m² and 125°C respectively. The weight of steam condensed after throttling was 1·8 kg and the weight of water trapped in the separator was 0·2 kg. Estimate the dryness fraction of the steam in the main.

(Ans. 0·882)
[U.E.I.]

22. A quantity of steam at a pressure of 0·9 MN/m² and 0·9 dry occupies a volume of 0·42 m³. It is expanded according to the law $pV^{1·35}$ = constant to a pressure of 0·3 MN/m².

Calculate:—

(a) the mass of steam present;
(b) the external work done;
(c) the change of internal energy;
(d) the heat interchange between the steam and surroundings, stating the direction of flow.

(Ans. (a) 2·17 kg; (b) 269 kJ;
(c) −890 kJ; (d) −621 kJ)
[U.E.I.]

23. (a) Explain the terms sensible heat, latent heat and total heat, as applied to a liquid and its vapour.

(b) Steam at 2 MN/m² and 300°C, is passed at constant rate through a vessel into which water is being sprayed, also at constant rate. If the water is initially at 15°C, determine the amount to be sprayed in per kg of steam entering the vessel, in order that the resultant mixture shall leave the vessel as dry saturated steam. Assume a constant pressure of 2 MN/m² for steam and water throughout the process, and neglect heat losses to surroundings.

(Ans. 0·0826 kg)
[U.L.C.I.]

24. Steam at a pressure of 1·2 MN/m² is passed into a heat exchanger where water at 0·15 MN/m² and 25°C is heated at constant pressure until its temperature is 15°C below saturation temperature. If the condensed steam leaves the heat exchanger at 50°C, determine the initial dryness fraction of the steam supply. Assume no heat losses to the surroundings and take the mass flow ratio water:·steam in the heat exchanger as 9:1.

If the steam supply is at the rate of 1200 kg/h determine the internal diameter of the steam inlet pipe to the heat exchanger in order that the velocity of steam through the pipe shall not exceed 10 m/s.

(Ans. 0·953; 81 mm)
[U.L.C.I.]

25. Define the total heat (or enthalpy) and the internal energy of a gas.

Show that the total heat of a fluid remains constant during a throttling operation. A sample of steam at 1·25 MN/m² is taken from a boiler and passed through a throttling calorimeter where, after throttling to 0·1 MN/m² its temperature is observed to be 110°C. Determine the dryness fraction of the steam leaving the boiler.

(Assume specific heat capacity of superheated steam is 2·1 kJ/kg deg C)

(Ans. 0·955)

[U.E.I.]

26. Describe the method of determination of the dryness fraction of steam using the combined separating and throttling calorimeter. Illustrate your description with simple sketches, including a line diagram of the arrangement of the apparatus. Explain the reason for using the separator.

Wet steam at 0·55 MN/m² is throttled to a pressure of 0·12 MN/m². Determine the lowest value of the dryness fraction of steam at 0·55 MN/m² which may be determined by the use of the throttling calorimeter alone.

(Ans. 0·968)

[U.L.C.I.]

27. One kilogramme of steam 0·9 dry at 1·2 MN/m² is expanded until its pressure is 0·12 MN/m². If the law of the expansion is $pV^{1·15} =$ constant, determine (a) the dryness fraction of the steam after expansion; (b) the work done during expansion; (c) the change of internal energy, and (d) the heat absorbed by or rejected from the steam during the operation.

(Ans. 0·762; 382 kJ/kg; −390 kJ/kg; −8 kJ)

[U.E.I.]

28. Describe, with the aid of sketches, the construction and operation of a combined separating and throttling calorimeter, when used for the accurate determination of the dryness fraction of steam.

The following readings were obtained when using the above apparatus: water collected in separator 35 g, steam condensed from throttling calorimeter 182 g, pressure of steam entering separator 1·1 MN/m², pressure and temperature after throttling 0·1 MN/m² and 112°C. Determine the true dryness of the steam entering the separator.

(Ans. 0·804)

[U.E.I.]

29. (a) Define enthalpy of steam and show that its values before and after a throttling process are equal.

(b) Steam leaves a boiler at 1·4 MN/m² and 200°C. After passing through a long pipeline without loss of pressure the steam enters an engine. A sample of this steam is passed through a throttling-calorimeter, where it is throttled to 0·1 MN/m² and 110°C. Determine (i) the dryness fraction of the steam entering the engine; (ii) the heat lost by the steam during its passage through the pipeline, expressed as a percentage of the initial total heat.

(Ans. 0·952; 3·82%)
[U.L.C.I.]

30. (a) 'A sudden reduction in the pressure of a vapour does not necessarily cause a change in the total heat of the vapour.' Explain this statement and show how it is used in determining the total heat of steam which is almost dry saturated.

(b) A low-pressure, hot-water heating system for a large building is provided by mixing exhaust steam 0·9 dry at 0·15 MN/m² with water at 18°C and 0·15 MN/m². Determine the steam: water ratio by weight if the resultant hot water is to leave the mixing vessel at (i) 52°C; (ii) 65°C.

(Ans. 0·0593; 0·0792)
[U.L.C.I.]

8. Steam Plant

The most efficient way of producing power from the energy released from a fuel when using steam as the working agent is demonstrated in a modern power station. Fig. 67 is a line diagram of such a station, and shows the main units required for the efficient conversion of the heat energy from coal fuel into electrical energy.

THE STEAM CIRCUIT

Water from the boiler drum flows through the steam tubes, where it is heated by the furnace gases and returns as wet steam to the drum. The wet steam then passes from the top of the boiler drum, through the nest of superheater tubes and returns to the main stop valve. Superheated steam is drawn off to the turbines where it expands, first in the high-pressure turbine and then in the low-pressure turbines, doing work on the blades of the turbines which drive the electrical generators.

After expansion, the steam exhausts from the low-pressure turbines to the condensers where it is condensed by an external supply of cooling water. The condensate from the condensers is pumped to the hotwell where any loss of water due to steam leakage is made up. The water from the hotwell is used as boiler feed, and is pumped into the boiler drum via the feedheaters and economizer. In the feedheaters the cold feedwater receives heat from partially expanded steam which has been bled from the turbines. Further heat is added to the feedwater in the economizer where the feedwater flows through tubes over which is passing the furnace gases which have been exhausted from the boiler, so that the feed enters the drum at or near the boiling temperature corresponding to the boiler pressure.

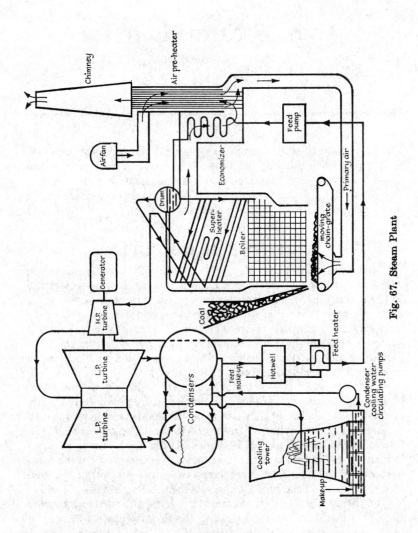

Fig. 67. Steam Plant

THE AIR/GAS CIRCUIT

A fan draws in air and forces it through a heater where it passes over plates heated on the other side by exhaust furnace gases on their way to the chimney. The hot air is ducted partly as primary air to the underside of the moving chain grate and partly as secondary air above the firebed. This air provides the oxygen necessary for the complete combustion of the coal supplied from the automatic stokers on to the firegrate. The hot products of combustion rise and circulate round the water tubes, usually being directed by baffles so that they pass over the tubes three times. On leaving the boiler, the gases pass over the tubes of the economizer and the plates of the air heater to be discharged at the chimney.

THE CONDENSER COOLING WATER CIRCUIT

A large supply of cooling water is required to extract the latent heat from the steam in order to condense it. When the power station is sited near a river, cooling water for the condensers is taken from the river and is pumped through the condenser tubes and finally discharged downstream. Where there is no such convenient supply of cooling water, cooling-towers are required, in which the water, warmed after its passage through the condenser tubes, may be cooled by the atmosphere. A good deal of cooling water is lost from the cooling-tower by its passing off as vapour with the air rising up the tower, so that make-up is continually added to the cooling-tower pond.

STEAM BOILERS

Boilers may be divided into two main types as follows:—

(a) *Water-tube boilers* in which water and steam flow through tubes surrounded by the furnace gases. This type is almost universally used in power station work because it permits of high working pressures, has a high rate of evaporation (up to 350 000 kg/h), and can be built in very large sizes with correspondingly high efficiency.

(b) *Fire-tube boilers* in which the hot furnace gases pass through tubes surrounded by water. This class includes Lancashire and locomotive boilers which operate at low and medium pressures (up to 1.5 MN/m²) and evaporate up to 7000 kg of steam/h.

LANCASHIRE BOILER

This fire-tube-type boiler is worthy of special mention because it is very widely used in industry.

Fig. 68 shows the general arrangement of the boiler. Two furnace tubes run the whole length of the boiler shell, and brick flues, which form an essential part of the boiler, redirect the gases underneath the shell and again along the outside of the shell before they are finally discharged to the chimney. In this way nearly all the external surface of the shell below the water line is heated in addition to the furnace tubes within the shell.

Wet steam is drawn off the top of the boiler and passes through superheater tubes which are placed at the end of the 'first pass', i.e. at the point where the furnace tubes discharge into the back flue.

BOILER EFFICIENCY

The efficiency of any boiler is defined by the relationship

$$\frac{\text{Heat transferred to feedwater in converting it to steam}}{\text{Heat released by complete combustion of the fuel}}$$

Suppose m_s kg of steam is formed when 1 kg of fuel is burnt.

Let h be the enthalpy of 1 kg of feedwater and

h_1 be the enthalpy per kg of steam raised.

Then for this flow process

∴ Heat transferred to 1 kg of feedwater in converting it to steam $= (h_1 - h)$

∴ Heat transferred to m_s kg of steam formed from 1 kg of fuel $= m_s(h_1 - h)$

Now heat released by complete combustion of 1 kg of fuel = Gross calorific value of the fuel.

$$\therefore \textbf{Boiler Efficiency} = \frac{m_s(h_1 - h)}{\textbf{Gross c.v.}}$$

A power-station water-tube boiler should have an efficiency of about 85%. A Lancashire boiler, with well-maintained brickwork and automatic stokers so that careful control of air supply is possible, should have an efficiency of about 75%.

Where does the remainder of the heat go? Since it is not employed usefully it is written down as a loss and is distributed as follows:—

196

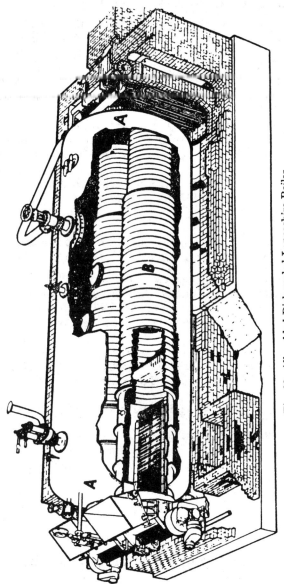

Fig. 68. All-welded Dish-ended Lancashire Boiler
A. Dish End   B. Corrugated Flue Sections

(John Thompson Ltd.)

1. Heat Loss to Chimney Gases

Since air and fuel are admitted to the boiler at the boiler-house temperature of say 20°C and are discharged from the chimney at say 200°C to 260°C, then some of the heat of the fuel has been used to heat them. Practical experience shows that the chimney-gas temperature should not be allowed to fall below 175°C so as to avoid condensation of acids formed from the combustion of small amounts of sulphur in the fuel.

The chimney gases are made up of (a) dry gases formed from the combustion of carbon and any sulphur present in the fuel, together with excess air supplied; (b) steam formed from the combustion of hydrogen together with any moisture initially present in the fuel. The heat loss to the chimney gases is divided between them as follows:—

(a) Heat loss to dry flue gas

For 1 kg of fuel burnt, let m_g kg of dry flue gas be formed.

For the flow process through the boiler furnace and chimney,

heat carried away by d.f.g. = increase of enthalpy of d.f.g.

$$\text{Heat loss to d.f.g.} = m_g c_p (t_2 - t_1) \text{ per kg of fuel burned.}$$

where t_2 = temperature of chimney gases.

t_1 = temperature of boiler house.

(b) Heat loss to steam in the flue gas

We assume the steam to exist as superheated steam at atmospheric pressure and chimney-gas temperature.

$$\text{Mass of steam formed per kg of fuel burned} = 9H_2 + \text{initial moisture per kg of fuel } (w)$$

Let $h_{\text{sup}} = $ enthalpy/kg of steam at atmospheric pressure and chimney-gas temperature.

h = enthalpy/kg of water at boiler-house conditions.

Then

$$\text{Heat loss to steam in the flue gas/kg of fuel burned} = (9H_2 + w)(h_{\text{sup}} - h)$$

2. Heat Loss to Unburnt Fuel falling through the Gratebars

When solid fuels are used, some of the fuel is always lost with the ash, usually because molten ash freezes round pieces of fuel and so prevents combustion. The weight of fuel lost is determined by heating ground-up samples of ash and measuring the reduction of weight.

3. Heat Loss due to Incomplete Combustion

Any CO present in the chimney gases is due to insufficient or inadequately distributed air supply. Since 1 kg of carbon burnt to CO releases only 10·5 MJ compared with 34·5 MJ when burnt to CO_2, the presence of CO in the flue gas represents a loss due to incomplete combustion.

4. Radiation Loss

Effective lagging is always necessary to reduce this to a minimum.

EQUIVALENT EVAPORATION 'FROM AND AT 100°C'

This definition refers to the mass of steam which a boiler can produce from the heat supplied by 1 kg of coal. In order to compare the evaporative capacity of different boilers, each of which may be producing steam at different pressures and temperatures, we suppose each boiler to be supplied with boiling feedwater at 100°C and to raise dry saturated steam at atmospheric pressure, i.e. at 100°C. Equivalent evaporation F and A refers to the mass of steam produced by a boiler from 1 kg of coal under these conditions. The equivalent evaporation F and A of a boiler can be deduced from its performance under normal running conditions as follows:—

Let steam be raised at pressure p and temperature t
h_1 = enthalpy/kg of steam at p and t
h = enthalpy/kg of feedwater supplied to boiler
then heat given to each 1 kg of steam raised = $h_1 - h$.

Now if the boiler were operating F and A 100°C, the amount of heat given to each kg of steam raised would be 2257 kJ, the latent heat of steam at 100°C.

Hence, for each kilogramme of steam raised at p and t, the boiler could raise $\dfrac{h_1 - h}{2257}$ kg of steam when operating F and A 100°C.

Therefore, if the boiler produces m_s kg of steam per kg of coal when operating at p and t,

Mass of steam raised per kg of coal when evaporating F and A 100°C $= \dfrac{m_s(h_1 - h)}{2258}$ kg

i.e. equivalent evaporation F and A 100°C $= \dfrac{m_s(h_1 - h)}{2258}$ kg

199

Example 47

'A Lancashire boiler is fired with coal of calorific value 30 MJ/kg and having an analysis of C 80%, H_2 5% and O_2 2%, the remainder being ash. Calculate the mass of air supplied per kg of coal if 30% in excess of that theoretically required is admitted, and find the heat lost to the dry flue gases per kg of coal given:—

flue gas temperature = 400°C, boiler-house temperature = 10°C, c_p for flue gas = 1·005 kJ/kg deg C.

Determine also the thermal efficiency and equivalent evaporation F and A of the boiler if 5000 kg of steam is raised/hour, dry saturated and at a pressure of 1·2 MN/m² from feedwater at 60°C, if 470 kg of coal are burned per hour.'

$$\text{Theoretical air supply} = \frac{100}{23}\left\{ \left(\frac{8}{3} \times 0\cdot 8 \right) + (8 \times 0\cdot 05) - 0\cdot 02 \right\}$$

$$= 10\cdot 9 \text{ kg}$$

$$\therefore \text{ Actual air supply} = 1\cdot 3 \times 10\cdot 9 = 14\cdot 2 \text{ kg}$$
$$\text{Excess air supply} = 3\cdot 3 \text{ kg}$$

Dry products/kg of coal:—

$$CO_2 = \left(\frac{11}{3} \times 0\cdot 8 \right) \qquad = 2\cdot 93$$

$$\text{Excess } O_2 = 0\cdot 23 \times 3\cdot 3 \qquad = 0\cdot 76$$
$$N_2 = 0\cdot 77 \times 14\cdot 2 \qquad = 10\cdot 93$$
$$\text{Total} \qquad \overline{14\cdot 62} = m_g$$

$$\therefore \text{ Heat lost to d.f.g./kg of coal} = m_g c_p (t_2 - t_1)$$
$$= 14\cdot 62 \times 1\cdot 005 (400 - 10) \text{ J}$$
$$= \underline{5\cdot 731 \text{ MJ}}$$

$$\text{Boiler efficiency} = \frac{m_s(h_1 - h)}{\text{Gross c.v.}}$$

$$\text{Mass of steam raised/kg of fuel } m_s = \frac{5000}{470} = 10\cdot 64 \text{ kg.}$$

$\therefore$ Boiler efficiency

$$= \frac{10\cdot 64(2784 - 4\cdot 187 \times 60) \times 10^3}{30 \times 10^6} = 0\cdot 898 \text{ or } 89\cdot 8\%$$

$$\text{Equivalent evaporation } F \text{ and } A = \frac{m_s(h_1 - h)}{2257} = \frac{10\cdot 64 \times 2531\cdot 5}{2258}$$

$$= \underline{11\cdot 93 \text{ kg/kg of coal}}$$

ECONOMIZERS

Feedwater is obtained by condensing the steam which is exhausted from the turbine. We may see from the steam tables, however, that its temperature is quite low because of the low pressure existing in the condenser (at 3 kN/m² the temperature of dry saturated steam is only 24·1°C). To save fuel, the feedwater should enter the boiler at as high a temperature as possible. The economizer does the job of heating the feedwater, using as its energy source the furnace gases after they have left the boiler and before they are discharged up the chimney.

AIR HEATERS

The air supplied to a furnace is preheated so as to reduce its cooling effect on the fire. The incoming air is drawn past plates heated on the other side by furnace gases passing from the economizer to the chimney.

CONDENSERS

The main purpose of a condenser is to create a region of low pressure into which the turbine or engine may exhaust. Fig. 69 compares the indicator diagrams of a condensing and a non-condensing engine, and illustrates the large increase of work obtained when a condenser is fitted.

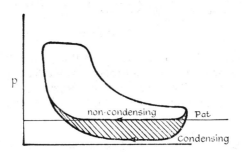

Fig. 69. Condenser

A secondary purpose of the condenser is to supply pure feedwater to the hotwell at a temperature nearly equal to that of the exhaust steam.

Condensation of the exhaust steam is brought about by extracting its latent heat using a flow of cold water. Upon condensation, the pressure of the steam is eliminated so that the pressure in the condenser falls considerably, often as low as 1.5 kN/m^2.

In order to maintain this low pressure, it is essential to extract any air which may enter the condenser chamber either by leakage or accompanying the exhaust steam. Air ejectors are therefore fitted and are operated by displacement pumps or steam jets.

Since the pressure in the condenser chamber is less than the atmospheric pressure, the condensate formed from the exhaust steam must be pumped out, and condensate extraction pumps are fitted for this purpose.

There are two main types of condensers, classified according to the way in which the cooling water cools the exhaust steam.

1. The Jet Condenser

In this type of condenser, jets of cold water are directed into the exhaust steam, extracting its latent heat by intimate contact. The type shown in Fig. 70 reserves one jet for the air ejector passage so as to reduce the loss of steam through the air ejector. Reference to

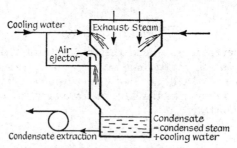

Fig. 70. Jet Condenser

the steam tables shows that to extract the latent heat of steam requires many times its weight of cooling water. The condensate is a mixture of the cooling water and the condensed steam, and cannot be used as boiler feed since the cooling water is normally impure.

2. The Surface Condenser

In this type of condenser, the cooling water and exhaust steam do not mix. A large number of tubes through which cooling water flows traverse the steam space of the condenser.

Heat transfer takes place across the tube walls, and the cooling water becomes warmed during the process of extracting the latent heat of the steam. Many arrangements of tubes have been used to obtain the best condensing conditions. For maximum efficiency, the vacuum should be as high as possible and the condensate temperature

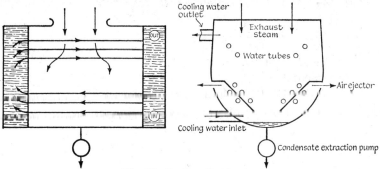

Fig. 71. Surface Condenser

should be as near as possible to that of the exhaust steam. For the first condition, air ejectors must be used and they are fitted at ports so placed that the air and associated vapour pass over tubes through which the coldest water is flowing, and baffle plates are placed so that condensate is prevented from falling on to these coolest tubes. To meet the second condition, the tube arrangement is such that the exhaust steam is in contact with the condensate.

The condensate from a surface condenser is a valuable source of distilled feedwater for the boiler, and is therefore pumped to the hotwell. Notice that since there is a vacuum in the condenser, it is necessary to pump the condensate out. To calculate the flow of cooling water required:—

Heat flowing into cooling water = heat flowing out of steam

$$\therefore 4\cdot187 \ \dot{m}_w(t_{\text{out}} - t_{\text{in}}) = \dot{m}_s(h_2 - h_c)$$

where $\dot{m}_w$ = mass of cooling water/hour
t_{out} = outlet temperature of cooling water
t_{in} = inlet temperature of cooling water
$\dot{m}_s$ = mass of condensate formed/hour
= mass of exhaust steam/hour
h = enthalpy per kg of exhaust steam
h_c = enthalpy per kg of condensate.

Example 48

'The steam used by a turbine is 5000 kg/h and exhaust occurs into a condenser operating at 10 kN/m². The exhaust steam is dry saturated at entrance to the condenser, and the condensate is undercooled 5 deg C. If the inlet and outlet temperatures of the cooling water are 15°C and 25°C respectively, determine the mass of water flowing per hour. If this water is to be carried in 300 pipes at a mean velocity of 2 m/s, determine a suitable pipe diameter.'

Enthalpy of dry sat. steam at 10
kN/m² $= 2584 \cdot 8$ kJ/kg

Temperature of dry sat. steam at 10
kN/m² $= 45 \cdot 8°$C

∴ Temperature of condensate for 5°C
undercooling $= 45 \cdot 8 - 5 = 40 \cdot 8°$C

∴ Enthalpy of condensate $= 4 \cdot 187 \times 40 \cdot 8$

$h_c = 170 \cdot 8$ kJ/kg

Mass flow of condensate/h, $\dot{m}_s = 5000$ kg

From $\dot{m}_w \times 4 \cdot 187 (25 - 15) = 5000 (2584 \cdot 8 - 170 \cdot 8)$

∴ $\dot{m}_w = 288\ 250$ kg/h.

$$\text{Volume of water passing per second} = \frac{288\ 250}{3600 \times 10^3}$$

$$= 0 \cdot 0807 \text{ m}^3$$

$$\therefore \text{ Volume of water/s per pipe} = \frac{0 \cdot 0807}{300} \text{ m}^3$$

$$\text{Flow} = \text{area} \times \text{velocity}$$

$$\therefore \text{ if } d = \text{pipe diameter}$$

$$\frac{0 \cdot 0807}{300} = \frac{\pi d^2}{4} \times 2$$

$$\therefore \ d = 0 \cdot 013\ 03 \text{ m} = 13 \cdot 03 \text{ mm}.$$

In the condenser chamber is a mixture of air and steam, and we must now refer to experimental laws concerning such mixtures.

DALTON'S LAWS OF VAPOURS

1. The pressure exerted by a mixture of gases, or gases and vapours, is the sum of the individual pressures which each would exert if it occupied the space alone. This is known as 'Dalton's Law of Partial Pressures', and assumes that all the individual gases are at the same temperature as the mixture.

2. The saturation pressure of a vapour depends only on its temperature, and the mass of a vapour which is required to saturate a given volume is the same whether or not other gases or vapours are present.

These statements may be difficult to understand at first sight. We have to remember, however, that whilst in a solid or liquid phase the molecules of a substance are relatively close together, whereas when the substance becomes a gas the molecules are widely separated and their volume is very small compared with the spaces between them. Accordingly, when gases are mixed there is plenty of room for all the molecules to move independently of each other.

Example 49

'A volume of 100 m³ of air at 0·2 MN/m² is saturated with moisture at 54°C. It is compressed to 0·3 MN/m². and 54°C. The pressures quoted are the total pressures in the vessel. Find:

(*a*) the masses of air and moisture present,
(*b*) the final partial pressure of the steam,
(*c*) the mass of moisture deposited on compression.

R for Air = 287 J/kgK.

[I.Mech.E.]

(*a*) From tables, at 54°C partial pressure of steam:—

$$p_s = 15 \text{ kN/m}^2$$
$$\therefore \text{ since } p = p_a + p_s$$

Initial partial pressure of air $p_{a1} = 200 - 15 = 185 \text{ kN/m}^2$.

For air, $\dfrac{pV}{T} = mR$

$$\text{Hence mass of air } m_a = \frac{pV}{RT}$$

$$= \frac{185 \times 10^3 \times 100}{287 \times 327}$$

$$= \underline{197 \cdot 2 \text{ kg}}$$

At 15 kN/m², specific volume of steam $V_g = 10 \cdot 02 \text{ m}^3/\text{kg}$

$$\therefore \text{ Mass of steam } m_{s1} = \frac{100}{10 \cdot 02} = \underline{9 \cdot 979 \text{ kg}}.$$

(*b*) At 54°C

$$p_s = \underline{15 \text{ kN/m}^2}$$

(*c*) Partial pressure of air $p_{a2} = 300 - 15 = 285 \text{ kN/m}^2$

Volume of air after compression $V_{a2} = \dfrac{mRT}{p}$

$$= \frac{197 \cdot 2 \times 287 \times 327}{285 \times 10^3}$$

$$= \underline{64 \cdot 92 \ m^3}$$

and this is the volume of the whole space and hence the volume occupied by the steam.

At 15 kN/m², specific volume of steam $V_g = 10 \cdot 02$ m³/kg

$$\therefore \text{ Mass of steam } m_{s2} = \frac{64 \cdot 92}{10 \cdot 02} = 6 \cdot 479 \text{ kg}$$

Hence mass of steam condensed on compression

$$= 9 \cdot 979 - 6 \cdot 479$$

$$= 3 \cdot 5 \text{ kg}$$

In a condenser, the combined pressure of the steam and leakage air is measured by the vacuum gauge. The partial pressure of the steam may be read from the steam tables if the temperature in the condenser is known. By difference then, the partial pressure of the steam may be obtained. Also, by law 2, the air and the steam in the condenser each may be considered to occupy the same volume, i.e. the volume of the condenser chamber.

Example 50

'Steam from a turbine exhausts into a surface condenser where the vacuum is 685 mm of mercury when the barometer reads 760 mm of mercury. The temperature at the air extraction outlet is 39°C. Determine the volume per kg of the air extracted, and the amount of vapour lost through the air ejector with each kilogramme of air extracted.'

$$\text{Pressure in condenser} = (760 - 685) \times 0 \cdot 133 \text{ kN/m}^2$$
$$= 10 \text{ kN/m}^2$$

At 39°C, partial pressure of steam at air extraction outlet
$$= 7 \text{ kN/m}^2 \text{ (steam tables)}$$

$\therefore$ Partial pressure of air at air extraction outlet
$$= 10 - 7$$
$$= 3 \text{ kN/m}^2$$

$$\text{For air } \frac{pV}{T} = mR \quad \therefore V = \frac{mRT}{p}$$

$\therefore$ for 1 kg of air at 3 kN/m² and 39°C

$$V = \frac{1 \times 287 \times 312}{3 \times 10^3}$$

$$= 29 \cdot 85 \text{ m}^3/\text{kg}$$

Since air and steam may be considered to occupy the same volume

$$\begin{array}{c}\text{Mass of air} \\ \text{present}\end{array} \times \text{volume/kg of air} = \begin{array}{c}\text{mass of} \\ \text{steam present}\end{array} \times \begin{array}{c}\text{volume/kg} \\ \text{of steam.}\end{array}$$

Now volume/kg of steam at $39°C = 20\cdot53$ m³/kg

$\therefore$ For each 1 kg of air, $1 \times 29\cdot85 = m_s \times 20\cdot53$

$\therefore$ Mass of steam associated with
each kg of air $\qquad m_s = \dfrac{29\cdot85}{20\cdot53}$

$$= 1\cdot454 \text{ kg}$$

Example 51

'Waste gases from blast furnaces are used in a steam generating plant consisting of boiler, superheater and economizer in series. The gases available amount to 100 000 kg/h at 760°C. The plant is designed for an efficiency of 70% when the atmospheric temperature is 15°C and the feedwater temperature is 35°C; the steam is generated at 0·5 MN/m². and 175°C. The heating surfaces are so proportioned that the boiler transmits 0·829 of the heat, the superheater 0·053 and the economizer 0·118. Calculate, taking c_p for gas $= 1100$ J/kgK,

(a) the mass of steam produced/hour;
(b) the temperature of the gases   (i) leaving the boiler
                                 (ii) leaving the superheater
                                 (iii) leaving the economizer
(c) the temperature of the water leaving the economizer;
(d) the dryness fraction of steam leaving the boiler.'

[U.E.I.]

(a) For whole plant, heat available/h $= 100\ 000 \times 1100(760 - 15)$ J
$\qquad\qquad\qquad\qquad\qquad\qquad\qquad = 81\cdot95$ GJ

$$\text{thermal efficiency} = \frac{\text{heat transferred to steam/h}}{\text{heat available/h}}$$

and heat transferred to steam/h $=$ mass of steam/h $\times (h_1 - h)$

$$\therefore\ 0\cdot7 = \frac{\text{mass of steam/h } (2800 - 4\cdot187 \times 35) \times 10^3}{81\cdot95 \times 10^9}$$

$\therefore$ mass of steam/h $= 21\ 650$ kg

207

(b) Heat trasnferred from furnace gases/h
= enthalpy gained by steam/h

(i) at the boiler

$$\dot{m}_{gas} \times c_p(t_{in} - t_{out}) = 0.829 \times \text{enthalpy gained by steam in plant/h}$$

$$\therefore 100\,000 \times 1100(760 - t_{out}) = 0.829 \times 0.7 \times 81.95 \times 10^9$$
$$\therefore t_{out} = \underline{326.7°C}$$

(ii) At the superheater
$$100\,000 \times 1100(326.7 - t_{out}) = 0.053 \times 0.7 \times 81.95 \times 10^9$$
$$\therefore t_{out} = \underline{299.1°C}$$

(iii) At the economizer
$$100\,000 \times 1100(299.1 - t_{out}) = 0.118 \times 0.7 \times 81.95 \times 10^9$$
$$\therefore t_{out} = \underline{237.5°C}$$

(c) At the economizer
Heat transferred from gases = enthalpy gained by feedwater
$$\therefore 100\,000 \times 1100(299.1 - 237.5) = 21\,650 \times 418.7(t_2 - 35)$$
$$\therefore t_2 = \underline{109.7°C}$$

(d) At the boiler

Enthalpy gained by feed-
water entering boiler/h = heat transferred from gases
being converted to wet steam passing over boiler/h

$$\dot{m}_s \{(h_{f1} + x_1 h_{fg1}) - h\} = \dot{m}_{gas} \times c_p(t_{in} - t_{out})$$
$$21\,650\{(640 + x \times 2109) - 4.187 \times 109.7)\} = 10^5 \times 1.1(760 - 326.7)$$
$$x = 0.957$$

REVISION EXERCISES—CHAPTER 8

1. During a boiler trial the following data were obtained: feed water 4500 kg/h at 20°C, coal 600 kg/h, boiler pressure 1·2 MN/m², steam condition 2·5% wet, c.v. of coal 30 MJ/kg. Calculate the boiler efficiency and the equivalent evaporation F and A 100°C per kg of coal. If an economizer was fitted and the feedwater heated to 110°C by the flue gases, determine the rate of coal consumption for the same rate of steam production assuming that boiler efficiency and steam condition remain as before.

(Ans. 66·2%, 10 kg, 514 kg/h)
[U.E.I.]

2. An 'Economic' boiler fitted with a mechanical stoker is supplied with coal having the following analysis—carbon 0·87, hydrogen 0·05, oxygen 0·04 and ash 0·04. Calculate the theoretical quantity of air required for complete combustion of 1 kg of fuel. When 40% excess air is supplied, the chimney-base temperature is 250°C and the boiler-house thermometer reads 20°C. Calculate the heat loss per hour to the dry flue gases when the boiler burns 370 kg of coal per hour of h.c.v. 30 MJ/kg. Find also the mass of dry saturated steam raised per hour from feedwater at 95°C to stop valve pressure of 0·6 MN/m² assuming a thermal efficiency for the boiler of 78%. Take c_p for dry flue gases = 1005 J/kg deg C.

(Ans. 11·65 kg, 1·438 GJ, 3669 kg)

3. Steam is raised at 1·5 MN/m² in a Lancashire boiler fitted with a superheater. Boiler feed is at a temperature of 45°C and the steam leaves the boiler and enters the superheater with a dryness fraction 0·97. The temperature of the steam leaving the superheater is 275°C. Find, taking c.v. of coal = 30 MJ/kg and boiler efficiency = 82%,

   (a) The mass of steam raised/h for a coal consumption of 500 kg/h.
   (b) The heat gained per kg of steam passing through the superheater.

(Ans. 4406 kg, 249·8 kJ)

4. The steam generated by a boiler is at a pressure of 1·5 MN/m² and 300°C. The feedwater temperature is 65°C. The boiler is oil fired and uses oil having a net c.v. of 44 MJ/kg. The ultimate analysis of the fuel is 85% C and 15% H_2. Find

   (a) The minimum weight of air required for complete combustion of the fuel.
   (b) The gross c.v. of the fuel.
   (c) The overall efficiency of the boiler if the weight of steam raised/kg of fuel is 12 kg.

Latent heat of water vapour 2454 kJ/kg.

(Ans. 15·12 kg, 47·31 MJ, 77·7%)
[L.U.]

5. 1300 kg of coal are burnt per hour in a boiler to produce 11 500 kg/h of steam at 1 MN/m² with a boiler efficiency of 67%. The water from the hotwell enters the economizer at 35°C and leaves at 110°C. For each kilogramme of coal fired 18 kg of air are supplied to the furnace at a temperature of 15°C. Take c.v. of coal = 30 MJ/kg, and c_p for flue gas = 1050 J/kg deg C, estimate

(a) The dryness fraction of steam leaving the boiler;
(b) The temperature of the flue gases
   (i) leaving the boiler;
   (ii) leaving the economizer.

(Ans. 0·978, 509°C, 370°C)

6. Describe with the aid of sketches either a Lancashire or a water-tube boiler.

A boiler burns 1350 kg of coal per hour of calorific value 30 MJ/kg. If 13 300 kg of dry steam at a pressure of 0·6 MN/m² is produced per hour from feedwater at 37°C, calculate the boiler efficiency.

(Ans. 85·4%)

7. In a trial of a water-tube boiler the following average readings were obtained:—
Analysis of coal supplied: C 85%, O_2 1·0%, H_2 4%, ash 10%.
Calorific value of coal = 30 MJ/kg.
Coal fired = 8600 kg/h. Mass of steam raised = 71 210 kg/h dry saturated.
Pressure of steam at boiler stop valve = 1·2 MN/m².
Feedwater temperature = 50°C. Temperature of flue gases at chimney base = 300°C. Boiler-house temperature = 20°C.
Flue gas analysis showed that 18·6 kg of air was supplied per kg of coal fired.
Take c_p for dry flue gas = 1·005 kJ/kg. Air contains 23% oxygen by weight. Estimate

(a) The percentage excess air supplied to the furnace;
(b) The mass of dry flue gas produced per kg of fuel burned;
(c) The heat lost to the dry flue gases per hour;
(d) The boiler efficiency.

(Ans. 66%, 19·14 kg, 4·63 GJ, 71%)

8. Explain why, in a surface condenser, a section of tubes is normally screened off so that the air is cooled to a relatively low temperature.

In such a condenser the pressure is 8 kN/m². The temperature at the screened section is 32·9°C. Find the mass of steam extracted/kg of air at the ejectors. Assume the pressure to be uniform throughout the condenser. Take R for air = 287 J/kgK.

(Ans. 1·04 kg)

9. Make sketches of two types of condenser in use with steam plant.

The vacuum gauge of a surface condenser reads 93 kN/m² of mercury when the barometric pressure is 100 kN/m². A thermometer in the steam space shows the temperature to be 29°C and

210

the volume of the steam space is 0·9 m³. Calculate the mass of air in the condenser.

Take R for air = 287 J/kgK.

(Ans. 0·031 34 kg)

10. The following results were obtained during a trial on a boiler plant employing an economizer and superheater. Total heating surface 300 m² made up as follows: economizer 70 m², boiler 140 m², superheater 90 m². Steam generated per hour 9040 kg. The feedwater enters the economizer at 20°C and leaves at 85°C. The steam leaving the boiler is at a pressure of 1·5 MN/m² and is of dryness fraction 0·95, and subsequently leaves the superheater at 225°C.

Coal used per hour 500 kg; Calorific value of coal 30 MJ/kg. Determine

(a) the rate at which heat is transmitted in each section of the plant;

(b) the efficiency of the plant.

(Ans. 15·3 MJ/m², 65·8 MJ/m², 7·33 MJ/m², 72·9%)

11. Explain why it is sometimes necessary to install a cooling tower in a steam plant using a surface condenser and state the function of the tower.

A two-pass surface condenser receives 8250 kg of steam per hour at 10 kN/m² and 0·92 dry. The inlet temperature of the cooling water is 30°C and the outlet temperature 35°C. The temperature of the hotwell is 45°C. The area of the cooling surface in the condenser is 120 m². The internal diameter of the tubes is 15 mm and the cooling water flows at the rate of 2 m/s. Determine

(a) the quantity of cooling water required in kg/s;

(b) the mass of steam condensed per hour per square metre of cooling surface;

(c) the number of tubes per pass.

(Ans. 241 kg/s; 68·75 kg, 683)

12. State briefly

(a) what are the main heat losses in a boiler plant;

(b) what measures are taken to minimize the losses.

The following are the mean values of the observations made during a boiler trial.

Coal used/hour = 60 kg; c.v. of coal = 34 MJ/kg

Feedwater pumped into boiler/hour = 500 kg

211

Temperature of feedwater $= 65°C$
Pressure in boiler $= 0·9$ MN/m^2.

A throttling calorimeter was used to measure the dryness fraction of the steam leaving the boiler and the steam after throttling was at a pressure of 0·1 MN/m^2 and temperature 105·6°C. Take c_p for superheated steam $= 1·9$ kJ/kgK. Estimate

(a) the dryness fraction of the steam leaving the boiler;
(b) the efficiency of the boiler.

(Ans. 0·962; 59·2%)

13. A boiler receives feedwater at 55°C and delivers 35 000 kg per hour of steam at 2 MN/m^2 at 275°C. The fuel consumption is 3400 kg having a calorific value of 35 MJ/kg. Calculate

(a) the equivalent evaporation from and at 100°C;
(b) the efficiency of the boiler;
(c) the velocity of steam in a 150 mm diameter delivery pipe.

(Ans. 12·47 kg, 80·8%, 65·25 m/s)

14. Sketch a line-diagram for the water-steam-water circuit in a modern steam-power plant, indicating the relative positions of the following units: superheater, economizer, feed heater, condenser, boiler, hotwell, prime-mover.

During a test of a steam-power plant, the following results were obtained:—

Temperature of water entering feed-heater $= 16°C$
Temperature of water leaving feed-heater $= 100°C$
Temperature of water leaving economizer $= 180°C$
Boiler pressure $= 2$ MN/m^2
Dryness fraction of steam leaving boiler $= 0·96$
Temperature of steam leaving superheater $= 325°C$

Calculate (a) the total heat gained by 1 kg of water; (b) the percentages of this total which are gained in the feed-heater, economizer, boiler, and superheater, respectively.

(Ans. (a) 3016 kJ, (b) 12%, 11%, 65%, 13%)
[U.L.C.I.]

15. A boiler has an evaporation of 40 000 kg/h from feedwater at 45°C, the steam being delivered from the superheater outlet at 1·9 MN/m^2 and 300°C. The steam entering the superheater at 1·9 MN/m^2 has dryness fraction 0·95. Calculate the heat gained in the superheater, expressed as a percentage of the total gain of heat between the feed inlet and the superheater outlet.

If the boiler efficiency is 76% when fired with oil of calorific value 43 MJ/kg, calculate the oil consumption in kg per hour.

(Ans. 11·4%, 3475 kg/h)
[U.L.C.I.]

16. Describe with the aid of sketches two types of condenser.

9000 kg of steam of dryness fraction 0·9 are condensed per hour in a jet condensor. It maintains a vacuum of 03 kN/m² when supplied with cooling water at 7°C. If the quantity of injection water supplied per minute is 4000 kg, calculate the rise in temperature of the cooling water. Barometer 760 mm mercury.

(Ans. 26·9 deg C.)

17. Distinguish between 'water-tube' and 'fire-tube' boilers and state under what circumstances each type would be used. Give details of methods used to increase the efficiency of modern water-tube boilers.

A boiler working at a pressure of 1·2 MN/m² evaporates 8 kg of water per kg of coal fired from feed water entering at 50°C. The steam at the stop valve is 0·95 dry. Determine the equivalent evaporation from and at 100°C.

(Ans. 8·77)
[U.E.I.]

18. A steam boiler plant includes a superheater and an economiser. Feed-water enters the economiser at 55°C and leaves at 150°C. In the boiler the steam is generated at 1·2 MN/m² and it leaves the superheater at a temperature of 300°C.

Determine the fuel consumption in kilogrammes per hour, using oil of calorific value 42 MJ/kg, when the boiler is steaming at the rate of 2500 kg/h, with an overall efficiency of 78%.

If the feed-water is returned to the boiler without passing through the economiser, determine the percentage increase in oil consumption.

What is the equivalent evaporation 'from and at 100°C' per kilogramme of fuel?

(Ans. 184·4 kg/h; 16·46%; 14·51 kg)
[U.L.C.I.]

19. A boiler generates 15 000 kg of steam per hour 0·95 dry at 0·7 MN/m² from feed-water at 20°C. The fuel consumption is 2400 kg/h of coal of calorific value 26 MJ/kg. Calculate the efficiency of this boiler, and its equivalent evaporation 'from and at 100°C', expressed as kilogrammes of steam/kg of coal.

The steam is used by a turbine developing 4000 kW. Calculate the overall efficiency of the plant.

(Ans. 61·9; 7·13 kg/kg; 23·1%)

[U.E.I.]

20. A coal-fired boiler produces 60 000 kg/h of steam at 4 MN/m² and 375°C from feed-water at 60°C. Determine the fuel consumption in kilogrammes per hour if the boiler efficiency is 88% when using coal of calorific value 27 MJ/kg.

What is the equivalent evaporation 'from and at 100°C' per kilogramme of fuel?

Determine the percentage increase in fuel consumption which would be necessary for the boiler to operate at the above conditions if the feed-water was returned to the boiler at only 15°C.

(Ans. 7330 kg/h; 10·52 kg/kg; 6·48%)

[U.L.C.I.]

21. (a) Describe, with the aid of sketches, the main features of a multi-tube boiler showing the path of the flue gases.

(b) During a test of a steam plant, the following readings were recorded: hot-well temperature, 18°C; temperature of water leaving feed-heater, 95°C; temperature of water leaving economiser, 165°C; boiler S.V. pressure, 2 MN/m²; dryness of steam at boiler stop valve, 0·97; temperature of steam leaving super-heater, 275°C.

Calculate (i) the total heat gained by 1 kg of feed-water during its passage through the plant; (ii) the percentages of this total which are gained in the feed-heater, economiser, boiler, and superheater respectively.

(Ans. 2890 kJ; 11·2%, 10·1%, 71%, 7·7%)

[U.L.C.I.]

9. The Steam Engine and the Steam Turbine

From 1769, when Watt invented the condenser, until recent years, the reciprocating steam engine was an important power producer. However, since 1930, the steam turbine has taken its place for stationary units and the 1956 plan for the use of electric and diesel traction on the railways spelt the doom of the reciprocating steam engine in this country. The main reasons for the replacement of the steam engine by the steam turbine are:—

1. The best steam engine efficiency obtained in practice was about 16% compared with a turbine efficiency of over 30%. This difference lies mainly in the ability to take advantage of good condenser vacuum. In a reciprocating engine a very large stroke is required to provide the large expansion needed and much of the resulting work is used to overcome friction. A turbine is at its best when provided with good vacuum, and mechanical friction losses are very small.
2. Since the steam flows through the turbine and is not held in a closed cylinder, as in an engine, turbines are compact and may be built to produce enormous amounts of power.
3. A turbine is a simple piece of machinery compared with a steam engine.

This section of the book therefore concentrates on the steam turbine and the references in previous editions to factors influencing the design of reciprocating steam engines have been omitted. Experience in carrying out energy balance accounts of reciprocating steam engines is, however, part of the programme for many students in colleges and a brief reference is given below.

INDICATED POWER

If the cut-off at each end of a double-acting engine is the same, the i.p. developed on the crank side of the piston will be less than that developed on the cover side because the crank-end effective

215

piston area is reduced by the area of the piston rod. Valve adjustment is sometimes made to allow for this and balance the power developed, but in any case it is desirable to calculate the i.p. developed at each end separately, thus:—

$$\begin{array}{ccc}\text{i.p.} & = \text{i.p.} & + & \text{i.p.}\\ \text{engine} & \text{cover} & & \text{crank}\end{array}$$

$$P_m \text{ (cover)} \frac{\pi}{4} D^2 L.n. + P_m \text{ (crank)} \frac{\pi}{4} (D^2 - d^2) L.n.$$

where D = piston diam.

d = piston-rod diam.

n = workings worked per second.

For multicylinder engines, the i.h.p. of the engine is the sum of the i.p. developed in the separate cylinders.

THERMAL EFFICIENCY

The indicated thermal efficiency of a steam engine is given by the relationship:—

$$\eta_t = \frac{\text{work done per minute}}{\text{heat input per minute}}$$

$$\eta t = \frac{\text{indicated power in Watts}}{\text{steam consumption kg/s } (h_1 - h)}$$

where h_1 = enthalpy/kg of steam supply (J/ks)

h = enthalpy/kg of condensate (J/kg)

From this expression we may write:—

$$\frac{\text{Steam consumption kg/h}}{\text{i.p. (kW)}} = \frac{3600 \times 10^3}{\eta_t (h_1 - h)}$$

i.e. specific steam consumption kg/kwh $= \dfrac{3600 \times 10^3}{\text{work done/kg of steam}}$ (ind.) supplied

ENERGY BALANCE ACCOUNT

By far the greatest proportion of the energy in the steam supplied to a steam engine is lost to the condenser. This is because most of the energy in the steam is latent heat which is removed by the condenser cooling water to bring about condensation. Useful energy remains in the condensate since this may be used as boiler feedwater.

I. Energy supplied per minute (measured above 0°C)

= Mass of steam used per min × enthalpy/kg of supply steam

$= m_s h_1$

II. Energy distributed per minute:—

THE STEAM ENGINE AND THE STEAM TURBINE

(a) To useful work = brake power × 60
(b) Energy transferred to condenser cooling water
$$= \dot{m}_w \times \text{specific heat capacity} \times (t_0 - t_1)$$
where $\dot{m}_w$ = cooling water flow lb/min
$t_0 = $ „ „ outlet temperature.
$t_1 = $ „ „ inlet temperature.
(c) Energy remaining in condensate
$$= \dot{m}_s \times \text{specific heat capacity } (t_c - 39)$$
where $\dot{m}_s$ = mass of condensate kg/min
= mass of steam supply kg/min
t_c = temperature of condensate
(d) Heat flow to surroundings by radiation, leakage etc. (by difference).

Example 52
'The following readings were taken during a trial of a single-cylinder double-acting steam engine:—

Bore 600 mm   Piston rod diam. 90 mm
Stroke 900 mm   Speed 158 rev/min
Barometer 765 mm   Condenser Vacuum 705 mm
Mean effective pressure (cover end) = 252 kN/m²
    „   „   „   (crank end) = 258 kN/m²
Condensate temperature = 40°C
Steam supply pressure 900 kN/m²
Condenser cooling water flow 1550 kg/min
Inlet temperature of cooling water 15°C
Outlet   „   „   „   „   31°C
Steam supply is dry saturated
Mass of condensate collected = 2760 kg/hour
b.p. developed = 225 kW.

Calculate the thermal and mechanical efficiency of the engine and draw up an energy balance account on a one-minute basis.'

Piston area (cover) $= \dfrac{\pi}{4} \times 0\cdot6^2 = 0\cdot09\pi \text{ m}^2$

Piston area (crank) $= \dfrac{\pi}{4}(0\cdot6^2 - 0\cdot09^2) = 0\cdot088\pi \text{ m}^2$

$$\underset{\text{engine}}{\text{i.p.}} = \underset{\text{cover}}{\text{i.p.}} + \underset{\text{crank}}{\text{i.p.}}$$

$$= 252 \times 10^3 \times 0\cdot09\pi \times 0\cdot9 \times \frac{158}{60}$$

$$+ 258 \times 10^3 \times 0\cdot088\pi \times 0\cdot9 \times \frac{158}{60}\text{W}$$

$$= 337\cdot9 \text{ kW}$$

217

$$\text{Mechanical efficiency} = \frac{\text{b.p.}}{\text{i.p.}} = \frac{225}{337 \cdot 9} = 66 \cdot 6\%$$

$$\text{Indicated thermal efficiency} = \frac{\text{i.p.}}{\text{kg of steam/second } (h_1 - h)}$$

From steam tables

$$h_1 = 2772 \cdot 1 \text{ kJ/kg}$$
$$h_c = 40 \times 4 \cdot 187 \text{ kJ/kg.}$$

$\therefore$ Indicated thermal efficiency $= \dfrac{337 \cdot 9 \times 10^3 \times 60}{\dfrac{2760}{60}(2774 - 40 \times 4 \cdot 187) \times 10^3}$

$$= \underline{16 \cdot 9\%}$$

Energy Supplied per minute
Energy supplied/min

$$= \text{Mass of steam/min} \times h_1$$

$$= \frac{2760}{60} \times 2774 \times 10^3$$

$$= 135 \cdot 9 \text{ MJ}$$

Energy Expenditure per minute
(a) To useful work $= \text{b.p.} \times 60$
$$= 225 \times 10^3 \times 60 \text{ J}$$
$$= 13 \cdot 5 \text{ MJ}$$
(b) To condenser cooling water
$$= m_w(t_0 - t_1) \times 4187 \text{ J}$$
$$= 1550 (31 - 15) \, 4187$$
$$= 103 \cdot 9 \text{ MJ}$$
(c) To condensate
$$= m_s \times t_c \times 4187 \text{ J}$$

$$= \frac{2760}{60} \times 40 \times 4187 \text{ J}$$

$$= 7 \cdot 70 \text{ MJ}$$
(d) To radiation, leakage, etc., by difference $= 10 \cdot 8 \text{ MJ}$

Energy supplied/ min	MJ	%	Energy distributed/min	MJ
Energy in steam supplied	135·9	100	Useful work Condenser cooling water Remaining in condensate Radiation etc	13 5 103·9 7·7 10·8
	135·9	100		135·9

218

Example 53

'In a test on a single-cylinder double-acting steam engine, the following observations were made:—

i.p. 25·1 kW, b.p. 21·2 kW, Steam supply 500 kN/m² at 175°C, Condenser vacuum 600 mmHg, Barometer 750 mmHg, Steam consumption per hour 300 kg. Condensate temp. 40°C. Condenser circulating water 7500 kg/h, Inlet temp. 16°C, Outlet temp. 37°C

(*a*) Calculate the indicated thermal efficiency.

(*b*) Estimate the dryness of the steam entering the condenser.

(*c*) Draw up an energy balance account for the engine in kJ/min.

Enthalpy of supply steam = 2800 kJ/kg
Enthalpy of condensate = 40 × 4·187 = 167·5 kJ/kg
Condenser pressure = (750 − 600) × 133·3 = 20 000 N/m² = 20 kN/m²

(*a*) Indicated thermal efficiency $= \dfrac{\text{i.p.} \times 3600 \text{ (J)}}{\text{kg of steam/min } (h_1 - h)}$

$$= \frac{25{\cdot}1 \times 10^3 \times 3600}{300 \times (2800 - 167{\cdot}5)}$$

$$= \underline{0{\cdot}1144 \text{ or } 11{\cdot}44\%}$$

(*b*) Heat flow from exhaust steam/h = heat flow into condenser cooling water/h

$$m_s\{(h_{f_2} + x_2 h_{fg_2}) - h\} = m_w(t_0 - t_1) \times s_p \text{ Ht. Cap.}$$
$$300\{(251 + x \times 2358) - 167{\cdot}5\} = 7500 \times 4{\cdot}187(37 - 16)$$
$$\therefore \; \underline{x_2 = 0{\cdot}896.}$$

(*c*) Energy supplied/min

$$= \frac{300}{60} \times 2800 \qquad\qquad = 14\,000 \text{ kJ/min}$$

Energy distributed/min

(*a*) Useful work = 21·2 × 60   = 1272 kJ/min

(*b*) To condenser cooling water

$$= \frac{7500}{60} \times 4{\cdot}187 \times 21 \qquad = 10\,990 \text{ kJ/min}$$

(*c*) To condensate

$$= \frac{300}{60} \times 4{\cdot}187 \times 40 \qquad\quad = 837{\cdot}4 \text{ kJ/min}$$

(*d*) To Radiation etc. (by difference) = 900·6 kJ/min

THE STEAM TURBINE

The chief advantage of a steam turbine over a reciprocating engine is its ability to convert into useful work a good proportion of the energy available in the large volumes of low-pressure steam. In a reciprocating engine this steam must be rejected to exhaust, because it is not practicable to build the cylinder large enough to contain the steam as it expands to low condenser pressures.

A steam turbine consists essentially of two parts: the *nozzle* in which the internal energy of high-pressure steam is converted into kinetic energy so that the steam issues from the nozzle with very high velocity; and the *blades* which change the direction of the steam issuing from the nozzle so that a force acts on the blades and propels them.

Steam turbines are classified as IMPULSE or REACTION types depending on how the steam expansion takes place.

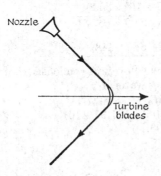

Nozzle

Turbine blades

Fig. 72. Impulse Turbine

In an IMPULSE turbine, all the available pressure drop from supply to exhaust occurs across the nozzles. The steam carries out its full expansion in the nozzles and emerges with high velocity. The nozzle is directed so that the steam glides on to the blades, and these are so shaped that the direction of flow of steam is changed and thereby a force is exerted on the blades.

In a REACTION turbine, only part of the pressure drop occurs in the nozzles, the remainder occurring during the passage of the steam through the blades. The blade passages are nozzle shaped so that the acceleration of the steam occurs partly in the nozzles, and partly in the blades. Since acceleration requires a force (force = mass × acceleration), a resultant reaction occurs on the blades. The force of reaction is added to the force resulting from re-direction of the steam to make the total propulsive force on the blades. Such a turbine would be better called an impulse-reaction type, since the propulsive force is due partly to each action. In most modern reaction turbines the propulsive force is derived from 50% reaction and 50% impulse effect.

In an impulse turbine nozzles are arranged in the turbine casing, whereas in the reaction turbine a ring of fixed blades acts as the nozzles for each stage.

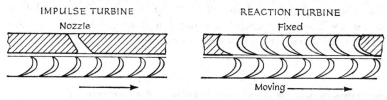

Fig. 73. Impulse and Reaction Turbines

THE SIMPLE IMPULSE TURBINE

The simplest case is that of a single-stage impulse turbine, i.e. one having one set of nozzles and a single ring of blades. The motive power of the turbine is obtained from the change of momentum of the jet of steam as its direction is changed during its flow over the blades of the rotor.

The flow of steam through a nozzle. The purpose of a nozzle is to convert the internal energy of the flow of steam into kinetic energy and this is done by expanding the steam from a higher to a lower pressure. The passage of steam through the nozzle occurs very quickly so that there is insufficient time for heat to flow to the surroundings and the expansion is therefore considered to be adiabatic.

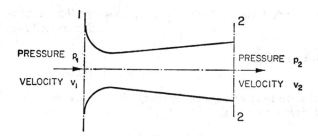

Fig. 74. Steam Flow through Nozzle

Consider the flow of steam initially at pressure p_1 and enthalpy H_1 expanding through a nozzle to a final pressure p_2 and enthalpy H_2. The steam is being conducted through a steady flow process, and applying the general energy equation to section 1.1 at entry and section 2.2 at exit we have

$$9 \cdot 81 \, mZ_1 + \frac{mv_1^2}{2} + H_1 + Q = 9 \cdot 81 mZ_2 + \frac{mv_2^2}{2} + H_2 + W$$

221

The change of potential energy is negligible, there is no heat flow across the boundary since the expansion is adiabatic, and no work flow occurs across the boundary, hence:

$$\frac{mv_2^2}{2} - \frac{mv_1^2}{2} = H_1 - H_2$$

In practice the inlet velocity of the steam is negligible compared with its exit velocity, and hence for 1 kg of steam we write

$$\frac{v_2^2}{2} = h_1 - h_2$$

$$\therefore \ v_2 = \sqrt{2(h_1 - h_2)}$$

It is necessary in practice to take account of the friction which occurs between the steam and the nozzle surface and within the steam stuff itself, the effect being to make the final steam drier and increase its specific enthalpy. Hence we may write

$$v_2 = \sqrt{2 \times \text{Actual enthalpy drop (joules)}}$$

Mass flow of steam. With a non-compressible fluid whose density remains approximately constant at different pressures (e.g. water), we have for flow through a pipe:—

Volume flowing/second (m³) = Area of cross-section of pipe (m²)
× flow velocity (m²/s)

$$= Av$$

$\therefore$ Mass flowing per second $= \rho Av$
where $\rho =$ density of fluid (kg/m³)

Steam, as we have seen, changes its specific volume considerably as the pressure alters, and

$$\rho = \text{density of steam (kg/m}^3)$$

$$= \frac{1}{V} \text{ where } V = \text{specific volume (m}^3\text{/kg)}$$

$\therefore$ Mass flow per second

$$\dot{m} = \frac{Av}{V} \text{ kg/s}$$

where $A =$ area of section considered (m²)
$v =$ velocity of steam at section (m/s)
$V =$ specific volume of steam at section (m³/kg)

Example 54

'Steam is admitted to a nozzle at 1 MN/m² dry saturated and expands adiabatically through the nozzle to a final pressure of 70 kN/m². The specific enthalpy of the leaving steam is 2279 kJ/kg and its dryness friction is 0·833. Find the velocity of the steam issuing from the nozzle, and the cross-sectional area required at exit of the nozzle for a flow of 0·75 kg/s.'

From tables, at 1 MN/m² $h_1 = 2778$ kJ/kg

Given $h_2 = 2278$ kJ/kg

∴ Enthalpy drop $h_1 - h_2 = 497 \cdot 2 \times 10^3$ J/kg

∴ Final steam velocity $v_2 = \sqrt{2 \times \text{Actual Enthalpy drop}}$

$$= \sqrt{2 \times 500 \times 10^3}$$

$$= 1000 \text{ m/s}$$

Specific volume of leaving steam at 70 kN/m² and 0·833 dry, from tables

$$V_2 = 0 \cdot 833 \times 2 \cdot 36$$
$$= 1 \cdot 967 \text{ m}^3/\text{kg}$$

Mass flow $\dot{m} = \dfrac{Av}{V}$

∴ Cross sectional at exit area $A_2 = \dfrac{\dot{m}V_2}{v_2}$

$$= \frac{0 \cdot 75 \times 1 \cdot 967}{1000}$$

$$= 0 \cdot 00\ 1475 \text{ m}^2$$

$$= \underline{1475 \text{ mm}^2}$$

The design of nozzles. The shape of the nozzle must be such that the conversion from internal energy to kinetic energy is carried out with greatest efficiency. There are two types of nozzle in use, converging and converging–diverging, and the selection of one or the other depends upon the pressure drop which occurs between inlet and outlet.

It is beyond the scope of this book to examine in detail the design of nozzles, but more advanced work shows that the area of cross-section of a nozzle decreases to a minimum at a pressure equal to about half the inlet pressure. This minimum section of a nozzle is called the throat, the corresponding pressure at the throat is called the

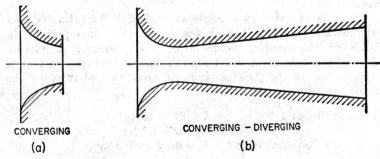

CONVERGING
(a)

CONVERGING – DIVERGING
(b)

Fig. 75. Nozzle Types

critical pressure, and it can be shown that the steam velocity at the throat is the same as the velocity of sound in steam. The type of nozzle required depends upon the discharge pressure, for if the discharge pressure is greater than the critical pressure a converging nozzle only is required, whereas if it is less than the critical pressure a converging–diverging nozzle is required.

In practice, the shape of a convergent–divergent nozzle is usually decided by calculating the throat and exit areas and making a simple conical section having an angle of divergence of 12°. The inlet to the throat is a smooth curve and is kept short in length (Fig. 76).

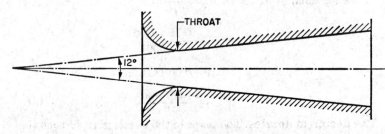

Fig. 76 Convergent–Divergent Nozzle

Example 55

'Steam at 1·20 MN/m² with a specific enthalpy of 2910 kJ/kg is expanded in a convergent–divergent nozzle to a back pressure of 100 kN/m².

At the throat the steam pressure is 660 kN/m², and the specific enthalpy 2785 kJ/kg, and at exit the specific enthalpy is 2505 kJ/kg. Determine the necessary throat and exit diameters of the nozzle for a discharge of 600 kg/h.'

At the throat:—

$$v_2 = \sqrt{2(h_1 - h_2)}$$
$$= \sqrt{2(2910 - 2785)}\,1000$$
$$= 500 \text{ m/s}.$$

We see from tables that the steam at the throat is still superheated

∴ Specific volume of steam at the throat

$$V_2 = \frac{230 \cdot 8(h - 1941)}{p}$$

$$V_2 = \frac{230 \cdot 8\,(2785 - 1941)}{660 \times 10^3}$$

$$= 0 \cdot 2951 \text{ m}^3/\text{kg}$$

Mass flow of steam $\dot{m} = \dfrac{A_2 v_2}{V_2}$

$$\therefore A_2 = \frac{600}{3600} \times \frac{0 \cdot 2951}{500} = \frac{0 \cdot 0984}{1000} \text{ m}^2$$

$$= 98 \cdot 4 \text{ mm}^2.$$

∴ Diameter at throat $= \sqrt{\dfrac{4 \times 98 \cdot 4}{\pi}}$

$$= 11 \cdot 2 \text{ mm}.$$

At exit:—

$$v_3 = \sqrt{2(h_1 - h_3)}$$
$$= \sqrt{2(2910 - 2505)}\,1000$$
$$= 900 \text{ m/s}.$$

For a final specific enthalpy of 2505 kJ/kg at 100 kN/m² final steam is wet and the final dryness fraction is given by:—

$$h_{\text{wet}} = h_f + x h_{fg}$$
$$\therefore 2505 = 417 \cdot 5 + x_3 \times 2257 \cdot 9$$
$$\therefore x_3 = 0 \cdot 927$$

∴ Specific volume of steam at exit

$$V_3 = x_3 V_g$$
$$= 0 \cdot 927 \times 1 \cdot 694$$
$$= 1 \cdot 567 \text{ m}^3/\text{kg}.$$

Mass flow of steam $\dot{m} = \dfrac{A_3 v_3}{V_3}$

$$\therefore A_3 = \frac{600}{3600} \times \frac{1 \cdot 567}{900} \times \frac{0 \cdot 29}{1000}$$

$$= 290 \text{ mm}^2$$

$$\therefore \text{ Diameter at exit} = \sqrt{\frac{4 \times 290}{\pi}}$$

$$= 192 \cdot 3 \text{ mm}$$

Blade Angles and Power developed in the Blades of Impulse Turbines. The blades are designed so that the steam from the nozzles will glide on to the blades without shock when they are running at the designed speed. If the angle of inclination of the nozzle to the blade ring $= a$, the speed of the steam jet from the nozzle $= v_1$ m/s, and the speed of the blade is u m/s, we can determine the correct inlet angle of the blade for no shock by drawing to scale the appropriate velocity diagram showing the distance moved in one second.

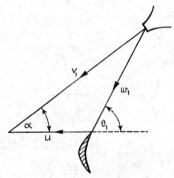

Fig. 77. Blade Angles

$v_1 =$ steam speed m/s

$u =$ blade speed m/s, i.e. peripheral velocity of turbine rotor.

$a =$ inclination of nozzles.

then $w_1 =$ velocity of steam jet relative to the moving blades at inlet (m/s).

and $\theta_1 =$ correct inlet-blade angle for no shock.

As the steam jet passes over the blade its direction is changed, and from Newton's second law of motion we have *'The rate of change of momentum is proportional to the externally applied force and takes*

226

place in the direction in which the force acts'. This law gives the equation for the force on the blades thus:—

Force on blades   =   Change of momentum per second
(in direction of motion)   (in direction of motion of the blades)

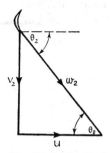

Fig. 78. Outlet-blade Angle

As the steam passes over the blade, friction reduces its velocity relative to the blade to w_2. For a given outlet-blade angle we can determine the actual velocity and direction of the leaving jet by drawing to scale the velocity diagram at exit from the blade (Fig. 78), thus:—

θ_2 = outlet angle of the blade. For cheapness in manufacture θ_2 is often made equal to θ_1 the inlet angle.

w_2 = velocity of steam jet relative to the moving blade at outlet (m/s). Note that w_2 is always *less* than w_1 in an impulse machine because of the friction loss.

u = blade speed (m/s).

then v_2 = final absolute velocity of the steam jet.

The complete velocity diagram can be shown as Fig. 84a or, more conveniently, as Fig. 84b in which the outlet velocity diagram has been moved up to use the same base u as in the inlet diagram.

Since, force on blades in   =   change of momentum per second
direction of motion   in direction of blade motion

for $\dot{m}_s$ kg of steam/s, $F = \dot{m}_s \left\{ \begin{array}{l} \text{horizontal component of} \\ \overleftarrow{v_1} \text{ in direction of motion} \\ -\text{horizontal component of} \\ \overleftarrow{v_2} \text{ in direction of motion} \end{array} \right.$

$$F = \dot{m}_s \left\{ v_{1u} - (-v_{2u}) \right\} \text{ Newtons}$$

(Notice that the component of v_2 in the direction of motion of the blades is negative.)

$$\therefore \ F = \dot{m}_s(v_{1u} + v_{2u}) \text{ Newtons}$$

227

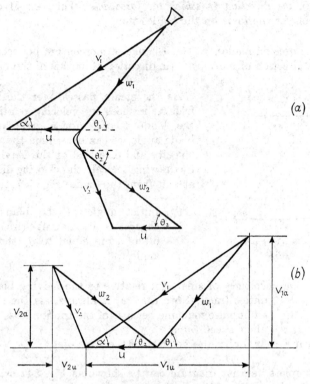

Fig. 79. Velocity Diagrams

Now work done on blades/s = force × distance moved/s

$\therefore$ work done/s $= \dot{m}_s(v_{1u} + v_{2u}) \times u$ (Newton metres = joules)

= power developed

Since the energy supplied to the blades per second is the kinetic energy of the steam jet $= \dfrac{\dot{m}_s v_1{}^2}{2}$:—

$$\text{Blading or diagram efficiency} = \frac{\text{work done on blades/s}}{\text{energy supplied/s}}$$

$$= \frac{\dot{m}_s(v_{1u} + v_2{}^n)\, u}{\dfrac{\dot{m}_s v_1{}^2}{2}}$$

228

$$\therefore \text{ blading or diagram} \atop \text{efficiency} = \frac{2u(v_{1u} + v_{1u})}{v_1{}^2}$$

It is now necessary to consider the change of momentum at right angles to the blades since any resulting force would cause an axial thrust.

$$\text{Force on blades perpendicular} \atop \text{to the direction of motion} = \dot{m}_s \left\{ \begin{array}{l} \text{vertical component of } v_1 \uparrow \\ -\text{vertical component of } v_2 \uparrow \end{array} \right\}$$

$$\therefore \text{ Axial thrust} = \dot{m}_s(v_{1a} - v_{2a}) \text{ Newtons}$$

$$\text{Energy converted to} \atop \text{heat by blade friction} = \text{loss of kinetic energy during} \atop \text{flow over blades}$$

$$= \frac{\dot{m}_s}{2}(w_1{}^2 - w_2{}^2)$$

Example 56

'A single-stage impulse rotor has a mean blade ring diameter of 0·6 m and rotates at a speed of 10 000 rev/min. The nozzles are inclined 20° to the direction of motion of the blades and the velocity of the issuing steam is 1200 m/s. Construct the velocity diagrams for the blades and determine the inlet angle to the blades in order that the steam shall enter the blade passages without shock. Assume a friction coefficient of the blading equal to 0·85 and that the inlet and outlet angles are equal. Find also

(a) the power developed at the blades for a steam supply of 3000 kg per hour;

(b) the diagram efficiency;

(c) the axial thrust;

(d) loss of kinetic energy due to blade friction.'

$$\text{Angular velocity of rotor} = \frac{10\ 000}{60} \times 2\pi \text{ radians/s}$$

$$\therefore \text{ Blade speed} \qquad u = \frac{2\pi \times 0\cdot3}{60} \times 10\ 000 \text{ m/s}$$

$$= 314\cdot2 \text{ m/s}$$

Steam speed $\qquad v_1 = 1200$ m/s

Steam flow $\qquad \dot{m}_s = 3000 \text{ kg/h} = \frac{3000}{3600} \text{ kg/s}$

From the diagram, drawing u and v_1, the construction gives

$$w_1 = 910 \text{ m/s}$$
$$\therefore \ w_2 = 0\cdot85 \times 910 = 775 \text{ m/s}$$
$$(v_{1u} + v_{2u}) = 1505 \text{ m/s}$$
$$(v_{1a} - v_{2a}) = 62 \text{ m/s}$$

229

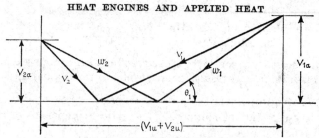

Fig. 80. Velocity Diagram

and by measurement:—

$$\theta_1 = 26\tfrac{3}{4}°$$

(a) power developed $\quad = \dot{m}_s(v_{1u} + v_{2u})u$

$$= \frac{3000}{3600} \times 1505 \times 314\cdot2 \text{ (W)}$$

$$= 394\cdot3 \text{ kW}$$

(b) Diagram efficiency $= \dfrac{2u(v_{1u} + v_{2u})}{v_1{}^2}$

$$= \frac{2 \times 314\cdot2 \times 1505}{1200 \times 1200}$$

$$= 0\cdot657 \text{ or } 65\cdot7\%$$

(c) Axial Thrust $\quad = \dot{m}_s(v_{1a} - v_{2a})$

$$= \frac{3000}{3600} \times 62$$

$$= 51\cdot7 \text{ N}$$

(d) Loss of K.E. due to blade friction

$$= \dot{m}_s(w_1{}^2 - w_2{}^2)$$

$$= \frac{3000}{3600} \times (910^2 - 775^2)$$

$$= 95\cdot9 \times 10^3 \text{ J} = 95\cdot9 \text{ kJ}$$

THE IMPULSE–REACTION TURBINE

The impulse–reaction turbine differs from the impulse turbine in that pressure drop occurs over the moving blades so that the steam expands continuously and its relative velocity at outlet is greater than its relative velocity at inlet. Fixed casing blades replace the nozzles of the impulse turbine, and pressure drop occurs both in the fixed casing blades and the moving blades. The exact opposite of

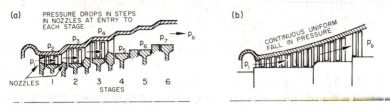

81. (a) Pressure-compounded impulse turbine (b) **Impulse reaction turbine**

the impulse turbine would be a full reaction turbine with all the pressure drop in the moving blades which would be so shaped that the passages between blades were of increasing cross-section. Pure reaction is impractical, and so-called reaction turbines are really mixed impulse–reaction turbines in which the drum is driven round partly by absorbing the kinetic energy of the steam emerging from the casing blades and partly by reaction from the pressure drop occurring in the moving blade passages (see Figs. 81 and 82).

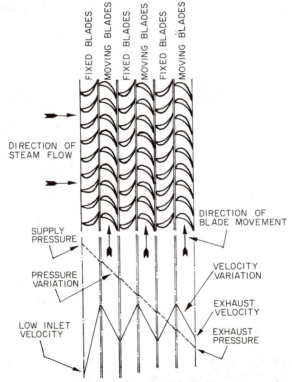

Fig 82. Group of reaction turbine blading—three pairs

The DEGREE OF REACTION is a measure of the proportion of the work done by reaction effect and may be defined as:

$$\frac{\text{Enthalpy drop in moving blade}}{\text{Enthalpy drop in moving blade} + \text{Enthalpy drop in fixed blade}}$$

In the Parsons impulse–reaction turbine, 'half-degree' or 50% reaction is used so that equal proportions of energy are contributed by the impulse and reaction effects. Thus the increase in relative velocity in the moving blades $(w_2 - w_1)$ exactly equals the increase in absolute velocity $(v_1 - v_2)$ in the casing blades. Hence the casing and moving blades must be made identical, and as a result the velocity triangle at outlet is identical in shape to that at inlet but is reversed in direction (Fig. 83).

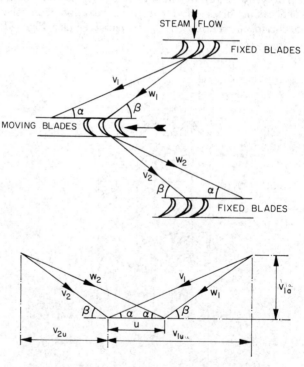

Fig. 83   Reaction blading—velocity triangles

Half the enthalpy drop which takes place in a reaction pair occurs in the casing ring, and the other half in the moving ring.

$$\frac{\text{power developed per reaction}}{\text{pair per kg of steam per second}} = (v_{1u} + v_{2u})u$$

With a symmetrical diagram of this kind it is better to obtain the solution mathematically rather than graphically. Thus,

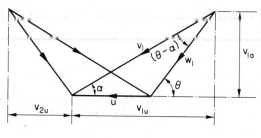

Fig. 84

Using the sine rule $\dfrac{u}{\sin(\theta - \alpha)} = \dfrac{v_1}{\sin(180 - \theta)} = \dfrac{w_1}{\sin \alpha}$

Also $v_{1u} + v_{2u} = v_1 \cos \alpha + w_1 \cos \theta$

and $v_{1a} = v_1 \sin \alpha$

It is now necessary to consider the height of the blades of a reaction turbine. In reaction turbines which 'flow full', the volume of steam flowing increases owing to the continual reduction of pressure, and the blade heights have to increase progressively to accommodate it. Steam fills the entire annular area between the base and shroud of the turbine blading apart from the area occupied by the blade section itself.

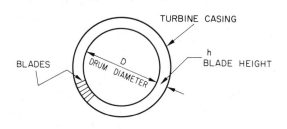

Fig. 85

Let D = diameter of the drum to the base of the blade (m)
and h = blade height (m)
Mean diameter of the blade circle = $(D + h)$ m

233

$\therefore$. Neglecting the area of the blades themselves, the annular area through which steam flows $= \pi h(D+h)$ m²

$$\left[\text{Alternatively } A = \frac{\pi(D+2h)^2}{4} - \frac{\pi D^2}{4} = \pi h(D+h)\right]$$

Let v_{1a} = axial velocity of flow of steam (m/s), as found from the velocity triangle.

Then the volume of steam flowing per second through the turbine $= \pi h(D+h)v_{1a}$ m³

At any reaction pair, the steam condition (x) can be found from the $h - s$ chart, and hence also the specific volume V

Then mass of steam flowing per second $= \dfrac{\pi(D+h)hv_{1a}}{V}$ kg

End thrust on the rotor = (Pressure drop per *moving* ring of blades × Blade area of the ring) + Any thrust due to v_a if the diagram is asymmetrical

Example 57

'A Parsons impulse–reaction turbine running at 400 rev/min with half-degree reaction develops 5000 kW and uses 6 kg of steam per kWh. The exit angle of the blades is 20° and the velocity of the steam relative to the blade at exit is 1·35 times the mean blade speed. At a particular stage in the expansion the pressure is 80 kN/m² and the corresponding dryness fraction of the steam is 0·95. Calculate for this stage a suitable blade height, assuming that the ratio

$$\frac{\text{Rotor diameter}}{\text{Blade height}} = 12.\text{'}$$

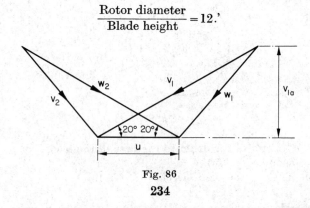

Fig. 86

234

$$w_2 = 1\cdot 35u = v_1$$

Axial velocity $v_{1a} = 1\cdot 35\ u\sin 20$
$$= 0\cdot 4616\ u$$

Now $u = \omega r$ and $\omega = \dfrac{2\pi}{60}\times 400,\ r = \dfrac{D+h}{2}$

where ω = angular velocity of the rotor

$$\therefore\ v_{1a} = 0\cdot 4616 \times \frac{2\pi}{60}\times 400 \times \frac{(D+h)}{2}$$

and $D = 12\ h$ $\therefore\ v_{1a} = 125\cdot 6\ h$ m/s ... (1)

Area of annulus $= \pi\ h(D+h)$
$$= 13\pi h^2 \qquad\qquad\qquad\ ...\,(2)$$

At 80 kN/m² and dryness 0·95

Specific volume of steam $= 0\cdot 95 \times 2\cdot 087 = 1\cdot 983$ m³/kg

$$\text{Flow} = \frac{5000 \times 6}{3600} = 8\cdot 33 \text{ kg/s}$$

$$\text{Flow} = \frac{\text{Area} \times \text{Velocity}}{\text{Specific volume}}$$

$$\therefore\ 8\cdot 33 = \frac{13\pi h^2 \times 125\cdot 6h}{1\cdot 983}$$

$$\therefore\ h = \sqrt[3]{\frac{8\cdot 33 \times 1\cdot 983}{13\pi \times 125\cdot 6}}$$

$$= \underline{0\cdot 1476 \text{ m}}$$

Example 58

'In a group of reaction turbine blading there are three fixed and three moving rings, all having the same blade section and through which the axial velocity of flow remains constant. The mean blade speed is 70 m/s.

For the mean moving blade ring the absolute and relative velocities at discharge are 30 and 90 m/s respectively, and the specific volume is 0·05 m³/kg.

Determine for a discharge of 3 kg/s of steam:

(*a*) the power developed by the group;

(*b*) the required area of the blade annulus;

(*c*) the enthalpy drop required by the group if the steam expands through it with an efficiency ratio of 0·80.'

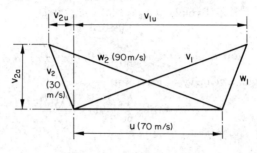

Fig. 87

(a) Power developed for the group $= 3(\dot{m}_s \times (v_{1u} + v_{2u}) \times u)$

$$= 3 \times 3(v_{1u} + v_{2u}) \times 70$$

$$\text{Now } 90^2 = v_{2a}{}^2 + (70 + v_{2u})^2 \qquad \dots (1)$$
$$\text{and } 30^2 = v_{2a}{}^2 + v_{2u}{}^2 \qquad \dots (2)$$
$$\text{Subtract } \therefore \ 90^2 - 30^2 = 70^2 + 2(70 v_{2u})$$
$$\therefore v_{2u} = 16 \cdot 43 \text{ m/s}$$
$$\text{then } v_{1u} + v_{2u} = 70 + (2 \times 16 \cdot 43)$$
$$= 102 \cdot 86 \text{ m/s}$$
$$\therefore \text{ power} = 3 \times 3 \times 102 \cdot 86 \times 70$$
$$= 64 \cdot 8 \text{ kW (for 3 pairs)}$$

(b) Blade annulus required (neglecting blade thickness)

$$= \frac{m_s \times \text{Specific volume}}{v_{2a}}$$

$$= \frac{3 \times 0 \cdot 05}{v_{2a}} \qquad \text{and} \qquad v_{2a} = \sqrt{30^2 - v_{2u}{}^2}$$

$$= \sqrt{900 - 16 \cdot 43^2}$$

$$= 25 \cdot 1 \text{ m/s}$$

$$= \frac{3 \times 0 \cdot 05}{25 \cdot 1}$$

$$= 0 \cdot 005 \ 975 \text{ m}^2$$

$$= 5975 \text{ mm}^2$$

(c) Enthalpy drop required $= \dfrac{\text{Total work done}}{0\cdot8 \times \text{Mass flow}}$

$$= \dfrac{64\cdot8 \times 10^3}{0\cdot8 \times 3}$$

$$= \underline{27 \text{ kJ/kg}}$$

REVISION EXERCISES—CHAPTER 9

1. During a test on a double-acting steam engine of bore 200 mm and stroke 300 mm loaded by a D.C. generator, the following readings were taken.

   Speed 125 rev/min
   Area of indicator card 700 mm²
   Spring No. 20 kN/m² per mm. Length of indicator card 65 mm
   Steam supply 300 kN/m² dry saturated
   d.c. generator volts 110, amperes 36, generator efficiency 0·85
   Condensate collected = 2·6 kg/min at 60°C
   Cooling water 55 kg/min, inlet temperature 17°C, outlet temperature 45°C.

   Calculate (a) i.p.; (b) b.p.; (c) mechanical efficiency; (d) indicated thermal efficiency, and draw up an energy balance account in kJ/min for the engine.

   (Ans. 8·46 kW, 4·66 kW, 55·1%, 7·89%.
   Work = 279·6, cooling 6449, condensate = 251·2, Radiation, etc., 105·2)

2. The following data were obtained from a test on a single-cylinder double-acting steam engine fitted with a rope brake.

   Cylinder diam. = 200 mm, Stroke = 250 mm
   Speed = 300 rev/min. Effective diameter of brake wheel = 800 mm
   Stop valve pressure = 750 kN/m². Dryness fraction of steam supply = 0·97
   Brake load = 150 kgf. Spring balance reading = 100 N.
   m.e.p. (from indicator cards) = 280 kN/m²
   Condenser pressure = 10 kN/m²
   Steam consumption = 3·7 kg/min
   Condenser cooling water = 90 kg/min
   Temperature rise of condenser cooling water = 14 deg C
   Temperature of condensate = 40°C.

Determine (a) b.p.; (b) i.p.; (c) mechanical efficiency; (d) indicated thermal efficiency; (e) steam consumption kg/kWh (brake); (f) energy balance account in kJ/min.

(Ans. 17·25 kW, 22 kW, 78·4%, 13·5%, 12·9 kg/kWh (brake), work = 1035 kJ/min, cooling = 5274 kJ/min, condensate = 167·5 kJ/min, Radiation, etc., 3523·5 kJ/min)

[U.E.I.]

3. A De Laval turbine has a wheel 150 mm mean diameter and runs at 24 000 rev/min. The nozzles are inclined at 20° to the plane of the wheel and the escape velocity of the steam from the nozzles is 900 m/s. There is a 10% loss of velocity in the blades and the inlet and outlet angles of the blades are equal. Determine:—

    (a) The blade angles;
    (b) the absolute velocity of the steam at the exit from the blades;
    (c) the wheel efficiency.

(Ans. $25\frac{1}{2}°$, 500 m/s, 58·5%)

4. The steam from the nozzles of a single-stage impulse turbine has a velocity of 850 m/s and is inclined at 20° to the direction of motion of the blades. Determine the necessary inlet angle of the blades so that no shock shall occur for a blade speed of 300 m/s.

Assuming that friction reduces the relative velocity of the steam by 10% as it passes over the blades and that the blade angles are equal, find the work done per kg of supply steam.

(Ans. 30°; 284·4 kJ)

5. Three nozzles direct the steam on to the blades of a De Laval turbine, the $\dfrac{u}{v_1}$ ratio of which is 0·4. The flow through each nozzle is 0·2 kg/s and the speed of the steam leaving the nozzle is 650 m/s. The inclination of the nozzles to the direction of blade motion is 16° and the blade inlet and outlet angles are equal. Determine the blade angles so that there shall be no shock at inlet, and allowing for a 10% reduction in the relative velocity over the blades, calculate the power developed by the turbine.

(Ans. 110 kW)

6. Steam issues from the nozzles of a single-stage impulse turbine at 900 m/s on to blades moving at 380 m/s. The blade-tip angles at inlet and exit are each 36°. The steam is to enter the blades without shock and the flow over the blades is frictionless.

Draw separate velocity diagrams for the passage of the steam to and from the blades. On the diagrams, indicate the absolute and relative velocities, including arrowheads to show the directions.

From the diagrams, or otherwise, determine (a) the angle at which the nozzles are inclined to the direction of motion of the blade; (b) the force exerted on the blades per kilogramme of steam per second; (c) the diagram efficiency.

(Ans. 22°, 910 N, 0·854)
[U.L.C.I.]

7. A single-cylinder, double-acting steam engine has a piston diameter of 250 mm and a stroke of 300 mm. The steam is admitted to the cylinder in a dry saturated state at 700 kN/m², and the exhaust pressure is 30 kN/m². Using a diagram factor of 0·82, calculate the indicated power when running at 160 rev/min with cut-off at 75% of the stroke. Neglect clearance and assume hyperbolic expansion.

What would be the alteration in steam consumption, expressed as kilogrammes per hour, if the cut-off was reduced to 30% stroke, all other cylinder conditions remaining unaltered?

(Ans. 41·7 kW, 466 kg/h)
[U.L.C.I.]

8. The nozzles of a single-stage impulse turbine are inclined at 22° to the plane of the wheel, and the blade inlet angle is 40°. If the velocity of the steam leaving the nozzles is 800 m/s, determine the value of the mean blade speed in order that the steam enters the blades without shock.

If the inlet and outlet angles of the blades are equal, and there is no loss due to friction as the steam flows over the blades, calculate the diagram efficiency.

Calculate also the steam consumption in kilogrammes per hour when 10 kW is being developed.

(Ans. 380 m/s, 85·5%, 132 kg/h)
[U.L.C.I.]

9. The nozzles of a simple impulse turbine are inclined at an angle of 25° to the plane of the moving blades. Steam leaves the nozzles at 850 m/s, the mean blade speed is 300 m/s, and the blade tip angles at inlet and exit are equal.

Assuming there is a 10% loss of velocity of steam during its passage through the blade-wheel, draw the inlet and exit velocity diagrams to a suitable scale. Hence determine (a) the

239

blade-tip angles; (b) the steam consumption in kilogrammes per second when the turbine is developing 150 kW; (c) the diagram efficiency.

(Ans. (a) 37°; (b) 0·66 kg/s; (c) 0·738)
[U.L.C.I.]

10. A single-stage impulse turbine has a nozzle angle of 22° and the blade inlet and outlet angles are equal. The mean blade speed is 330 m/s and steam leaves the nozzles at 500 m/s. Draw the inlet and outlet velocity diagrams, allowing for a friction factor of 0·88. Hence determine (a) the blade angles; (b) the power developed by a steam flow of 12 kg/min; (c) the diagram efficiency.

(Ans. 36°; 50·8 kW; 79·4%)
[U.L.C.I.]

11. A single-wheel impulse turbine has a rotor of mean diameter 600 mm rotating at 12 000 rev/min. The velocity of the steam leaving the nozzles is 2·8 times the mean blade speed, and the blade inlet and outlet angles are 35° and 27° respectively. The relative velocity of steam to blade at outlet is 0·78 of that at inlet.

Draw the inlet and outlet velocity diagrams; hence determine the diagram or blading efficiency, and the power developed by a flow of 700 kg of steam per hour.

(Ans. 74·2%; 80·5 kW)
[U.E.I.]

12. In a simple impulse steam turbine the nozzles are inclined at 22° to the plane of the blade-wheel, and steam leaves the nozzle at 1000 m/s. If the blade inlet angle is 34°, determine the mean velocity of the blades.

During its passage through the blade-wheel the steam loses 12% of its relative velocity due to friction; the blade outlet angle is 40°. Determine (a) the absolute velocity of the steam leaving the blade-wheel; (b) the diagram efficiency; (c) the steam consumption in kilogrammes per hour when the wheel is developing 20 kW.

(Ans. 350 m/s; 420 m/s; 74·2%; 194 kg/h)
[U.E.I.]

13. In a simple impulse steam turbine the nozzles are inclined at 20° to the plane of the blade-wheel. Steam leaves the nozzles at 950 m/s and the blade-wheel rotates at 20 000 rev/min. The inlet angle of the blade tips is 32°. Determine the blade speed and the mean diameter of the blade-wheel.

If the outlet angle of the blade tips is 35°, and 10% of the relative velocity of the steam is lost due to friction in the blade passages, determine the absolute outlet velocity of the steam. Steam flows through the blade-wheel at the rate of 680 kg/h. Determine the power developed and the blade-wheel or diagram efficiency.

For the velocity diagrams, use scale 1 cm = 50 m/s.

(Ans. 360 m/s; 343 mm; 335 m/s; 64 kW; 78·6%)

[U.L.C.I.]

14. A single-wheel impulse steam turbine has a mean wheel diameter of 0·35 m and runs at 24 000 rev/min. The nozzles are inclined at 23° to the plane of the wheel, and the steam leaves the nozzles at 1050 m/s. During its passage through the blade-wheel, the relative velocity of the steam is reduced by 12%. If the blade inlet and outlet angles are to be equal, determine (a) the blade angles; (b) the blade-wheel or diagram efficiency, and (c) the steam consumption in kg/h when the wheel is developing 20 kW.

(Ans. 38°; 77%; 169 kg/h)

[U.L.C.I.]

15. Steam is expanded adiabatically in a convergent–divergent nozzle from inlet conditions of 1 MN/m^2 and 275°C to a back pressure of 7 kN/m^2. The dryness fraction of the leaving steam is 0·9. Determine the exit area of the nozzle for a flow of 1·0 kg/s.

(Ans. 0·016 m^2)

16. Show that for the adiabatic expansion of steam through a nozzle the steam velocity at any section may be given by $v = 224\sqrt{h_1 - h}$ where h is the specific enthalpy of the steam at the section. State any assumptions made.

A De-Laval turbine has a single nozzle and one set of blades. If the adiabatic enthalpy drop of steam through the nozzle is 270 kJ/kg, find the velocity of the steam at exit from the nozzle.

(Ans. 734·8 m/s)

17. A single stage impulse turbine is fed by a set of nozzles through which the steam drops from 500 kN/m^2 and 350°C to a back pressure of 20 kN/m^2 dryness fraction 0·98. The steam is directed on to a single set of blades, the nozzle angle being 20° and the blade speed 300 m/s. Determine the necessary inlet angle of the blades so that no shock shall occur. Assuming that friction reduces the relative velocity of the steam by 10% as it passes over the blades, and that the blade angles are equal, find the work done per kg of supply steam.

[Ans. 31°]

18. Steam flows at the rate of 2 kg/s over the blades of a reaction turbine which has fixed and moving blade angles 35° and 20°. At a certain section the blade speed is 20 rev/s. Determine the power being developed at the section.

(Ans. 2746 kW)

19. In a reaction turbine the blade angles are 20° and 35° for both fixed and moving blades. If the drum diameter is 1 metre and the speed 360 rev/min, find the force on the blades and the power being developed per kg/s of steam flow.

(Ans. 59·66 N, 1·125 kW)

20. In a reaction turbine the blade angles are 35° and 20°. The guide blades are the same shape but reversed in direction. At a certain place the drum diameter is 0·9 metres and the blades are 100 mm high. The steam pressure is 200 kN/m² and its condition is 0·9 dry. If the speed of the turbine is 300 rev/min find the power developed in the ring of moving blades.

(Ans. 3·67 kW)

10. Introduction to Heat Transmission

Heat may flow in three different ways, by conduction, by convection or by radiation and we shall examine each of these methods in turn. It must be remembered that the process of heat transmission in practice is very complex, so that in most cases the process takes place by a combination of these methods, and design is usually accomplished as a result of a combination of experimental work and theoretical analysis.

CONDUCTION

The method of heat transmission by conduction lends itself fairly readily to mathematical treatment. The kinetic theory states that the absolute temperature of a body is proportional to the kinetic energy of its molecules. Thus, in a bar of metal which is being heated at one end so that a temperature gradient exists along the bar, the molecules of the metal at the hot end are moving faster than those at the cold end. Heat energy is transmitted from the hot to the cold end by impact between adjacent molecules, so that a continuous flow of energy occurs whilst the temperature gradient is maintained. If the heat source is removed from the hot end of the bar, the molecules gradually achieve equal energy by impact and the bar reaches a uniform temperature throughout. It is found experimentally that the quantity of heat transmitted per unit time by conduction is directly proportional to the cross-sectional area of the bar and the temperature gradient, and is inversely proportional to the length of the path. Thus, for heat transmission by conduction along a bar we may write:—

$$\dot{Q} = \frac{kA(\theta_2 - \theta_1)}{l}$$

where $\dot{Q}$ = quantity of heat transmitted per unit time W(J/s)
k = a constant known as the coefficient of thermal conductivity of the material. This coefficient is defined as 'the number of heat units flowing per unit time through unit cross-sectional area when the temperature falls by one degree through unit length of path'.

If $\dot{Q}$ is measured in W, temperature in degrees Centigrade and distances in metres, the units of k are

$$\frac{W}{m^2 \text{ deg C}/m}, \text{ or more usually } \frac{W}{m \text{ deg C}}$$

A = cross-sectional area of path, m²
$\theta_2 - \theta_1$ = temperature gradient, degC
l = path length, m

Conduction of Heat through Walls. Consider the transmission of heat through an area A of a wall of thickness l made of a material with a coefficient of thermal conductivity k and having surface temperatures θ_1 and θ_2 where $\theta_1 > \theta_2$

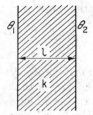 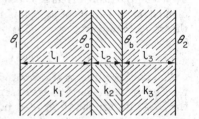

Fig. 88

Then heat transmitted per unit time $\dot{Q}$ is given by

$$\dot{Q} = \frac{kA(\theta_1 - \theta_2)}{l}$$

A wall may be composed of a number of slabs of different materials, each of thickness l_1, l_2, l_3, etc. and having coefficients of thermal conductivity k_1, k_2, k_3, etc. Let the temperatures at the wall surfaces be θ_1, θ_a, θ_b and θ_2 where $\theta_1 > \theta_2$ (as shown in Fig. 88). The heat transmitted from the hot side to the cold side is conducted through each slab in turn, so that the heat transmitted per unit area per unit time through each slab is the same, i.e.

$$\frac{\dot{Q}}{A} = \frac{k_1(\theta_1 - \theta_a)}{l_1} = \frac{k_2(\theta_a - \theta_b)}{l_2} = \frac{k_3(\theta_b - \theta_2)}{l_3}$$

Now $(\theta_1 - \theta_2) = (\theta_1 - \theta_a) + (\theta_a - \theta_b) + (\theta_b - \theta_2)$

$$= \frac{\dot{Q}l_1}{Ak_1} + \frac{\dot{Q}l_2}{Ak_2} + \frac{\dot{Q}l_3}{Ak_3}$$

$$= \frac{\dot{Q}}{A}\left\{ \frac{l_1}{k_1} + \frac{l_2}{k_0} + \frac{l_3}{k_0} \right\}$$

$$\therefore \dot{Q} = \frac{A(\theta_1 - \theta_2)}{\left\{ \dfrac{l_1}{k_1} + \dfrac{l_2}{k_2} + \dfrac{l_3}{k_3} \right\}}$$

Example 59

'A furnace wall consists of 230 mm of fire-brick and 115 mm of insulating-brick having thermal conductivities of 0·7 and 0·25 W/m deg C. Calculate the rate of heat loss per square metre of furnace wall when the temperature difference across the slab is 500 deg C.'

$$\dot{Q} = \frac{A(\theta_1 - \theta_2)}{\dfrac{l_1}{k_1} + \dfrac{l_2}{k_2}} \text{ W}$$

$$= \frac{500}{\dfrac{0·230}{0·7} + \dfrac{0·115}{0·25}}$$

$$= 634 \text{ W/m}^2$$

Conduction of Heat through Pipe Walls and Lagging. Fig. 89 represents the cross section of a pipe of radius r_1 which is lagged to radius r_2.

No significant error is introduced by neglecting the thickness of the metal wall of the pipe, and by assuming the temperature of the inner surface of the lagging θ_1 to be the same as that of the fluid passing through the pipe. The temperature of the outer surface of the lagging is θ_2, but it must be noticed that θ_2 is *not* the temperature of the surroundings. It is not easy to measure θ_2 accurately, but a good approximation can be obtained by attaching a thermometer to the outside of the lagging and shielding it with cotton-wool.

In order to calculate the heat conducted across the lagging, we must first observe that the area through which the heat is passing is varying, so that the method of the calculus must be used.

Consider the heat transmitted across a cylindrical element of the lagging at radius r and of thickness δr for a length of pipe L. Let the temperature gradient over the element be $\delta\theta$.

245

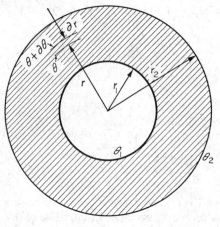

Fig. 89

Area through which heat is being transmitted:—
$$A = 2\pi r \times L$$
$$\text{Path length} = \delta r$$

Now the heat transmitted per unit time $\dot{Q}$ is directly proportional to the transmission area and to the temperature gradient, and is inversely proportional to the length of path

$$\therefore \dot{Q} \propto \frac{A \times (-\delta\theta)}{\delta r}$$

$$\text{or } \dot{Q} = k \times A \times \left(-\frac{\delta\theta}{\delta r} \right)$$

(A negative sign is given to $\frac{\delta\theta}{\delta r}$ in the equation because θ is reducing as r is increasing)

$$\therefore \dot{Q} = -k \times 2\pi r \times L \times \frac{\delta\theta}{\delta r}$$

Hence, in the limit

$$\dot{Q} \int_{r_1}^{r_2} \frac{dr}{r} = -k \times 2\pi L \int_{\theta_1}^{\theta_2} d\theta$$

$$\therefore \dot{Q} \log_e \frac{r_2}{r_1} = -k \times 2\pi L \, (\theta_2 - \theta_1)$$

$$\text{i.e. } \dot{Q} = \frac{k \times 2\pi L \, (\theta_1 - \theta_2)}{\log_e \dfrac{r_2}{r_1}}$$

Example 60
   'Steam at 260°C is passing through a 150 mm steam pipe lagged to

a diameter of 280 mm. The temperature of the surface area of the lagging is measured as 50°C. Taking the coefficient of thermal conductivity of the lagging as 0·6 W/m deg C, estimate the heat loss from 30 m length of pipe.'

$$\dot{Q} = \frac{k \times 2\pi L (\theta_1 - \theta_2)}{\log_e \frac{r_2}{r_1}} \text{ W}$$

$$= \frac{0\cdot6 \times 2\pi \times 30 \times (260 - 50)}{\log_e \frac{140}{75}}$$

$$= 38\cdot07 \text{ kW}$$

CONVECTION

The transmission of heat per unit time from a surface by convection is given by:—

$$\dot{Q} = hA(\theta_s - \theta_f)$$

where $\dot{Q}$ = heat transmitted per unit time
h = coefficient of convection heat transmission
θ_s = the temperature of the surface
θ_f = the temperature of the surrounding fluid

If the units of $\dot{Q}$ are W, of area are m², and of temperature difference degC, then the units of the coefficient h are W/m² degC.

The main problem facing the designer is to obtain appropriate values of the heat transmission coefficient h. It is necessary to differentiate at the beginning between NATURAL CONVECTION and FORCED CONVECTION. In the process of natural convection, the heat applied to a fluid causes it to expand and become less dense so that it rises away from the heat source and its place is taken by a colder and denser mass of the fluid. In this way natural 'convection currents' are set up and the heat energy is distributed. The term forced convection is applied when the movement of the fluid occurs because of outside forces; for example in stirring, or in the flow of fluid through a pipe which occurs because of a pressure head.

Determination of the coefficient h is extremely complicated since it may depend upon surface conditions, fluid velocity, viscosity, turbulence, density, specific heat and coefficient of expansion. To obtain experimentally a complete range of values of h taking account of all the variables and applying them to the whole range of liquids, vapours and gases which are in use in industry is such an enormous task as to be impractical. Results for specific sets of circumstances have of course been published, but these can be used with confidence only for identical conditions. A procedure known as 'dimensional analysis', which is beyond the scope of this book, can be used to make reasonable deductions for fluids of known characteristics operating in standard sections such as pipes and flat plates.

Example 61

'In a small furnace, the heat loss to the surroundings is to be kept down to 1700 W/m². The internal temperature of 150 mm fire-brick wall which lines the furnace is 650°C, and the temperature of the air surrounding the furnace is 25°C. Neglecting the temperature drop through the steel casing, calculate the thickness of exterior lagging required. Take the thermal conductivity of the fire-brick as 1 W/m degC, the thermal conductivity of the lagging as 1·2 W/m degC and the convection heat transfer coefficient of the lagging surface as 20 W/m² degC. Radiation from the lagging surface may be ignored. Estimate also the temperature of the outer surface of the lagging.'

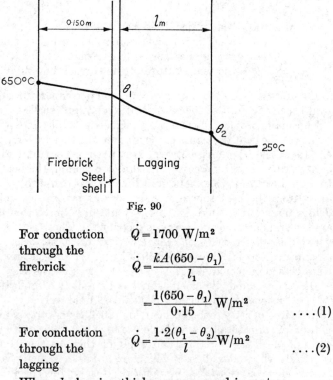

Fig. 90

For conduction through the firebrick

$$\dot{Q} = 1700 \text{ W/m}^2$$

$$\dot{Q} = \frac{kA(650 - \theta_1)}{l_1}$$

$$= \frac{1(650 - \theta_1)}{0·15} \text{ W/m}^2 \qquad \ldots (1)$$

For conduction through the lagging

$$\dot{Q} = \frac{1·2(\theta_1 - \theta_2)}{l} \text{W/m}^2 \qquad \ldots (2)$$

Where l = lagging thickness measured in metres

For convection
from the
lagging surface

$\dot{Q} = hA(\theta_2 - 25)$
$= 20(\theta_2 - 25) \ \text{W/m}^2$ ....(3)

From (1) $\quad 650 - \theta_1 = 1700 \times 0 \cdot 15$

From (2) $\quad \theta_1 - \theta_2 = 1700 \times \dfrac{l}{1 \cdot 2}$

From (3) $\quad \theta_2 - 25 = 1700 \times \dfrac{1}{20}$

Adding: $650 - 25 = 1700 \left(\dfrac{0 \cdot 15 + l + 0 \cdot 05}{1 \cdot 2} \right)$

$\therefore \dfrac{l}{1 \cdot 2} = \dfrac{625}{1700} - 0 \cdot 15 - 0 \cdot 05$

$\therefore \quad \underline{l = 0 \cdot 201 \ \text{m}}$

For the temperature of the lagging surface, from equation (3)
$$1700 = 20(\theta_2 - 25)$$
$$\therefore \ \theta_2 = \underline{110°C}$$

RADIATION

All bodies radiate energy in electromagnetic wave-forms which travel at a speed of 300 000 kilometres per second. The energy radiated is identified by its wavelength and particular bands of diminishing wavelengths are described as radio waves, infra-red rays, light rays, ultra-violet rays, X-rays and γ-rays.

We are particularly interested in thermal radiation which lies in the infra-red band extending over a wavelength from about $0 \cdot 8 \ \mu$ to $10 \ \mu$ (where $\mu = 10^{-3}$ mm). The thermal energy radiated by a body increases with its temperature, and the amount of thermal energy radiated can be measured by a thermopile which consists essentially of a sensitive thermocouple. If the extreme wavelengths of the energy received by the thermopile are λ and $\lambda + \delta \lambda$ and the total energy radiated from unit area of the source per unit time is E, the ratio $\dfrac{E}{\delta \lambda}$ is called the EMISSIVE POWER of the source E_λ. In Fig. 92 are plotted the values of E_λ for a black body at a variety of absolute temperatures. It can be seen from these curves that the energy radiated at each wavelength is increased as the temperature of the radiating black body is increased, and also that the wavelength corresponding to the maximum energy radiation becomes shorter as the temperature of the body increases. As the temperature of the

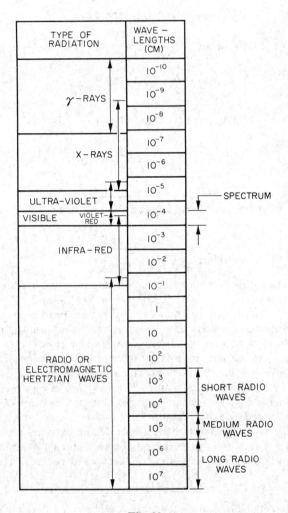

Fig. 91

body increases its radiation begins to include some of the visible spectrum, so that at about 550°C dull red appears, at about 870°C cherry red, at about 1200°C yellow, and white at about 1600°C.

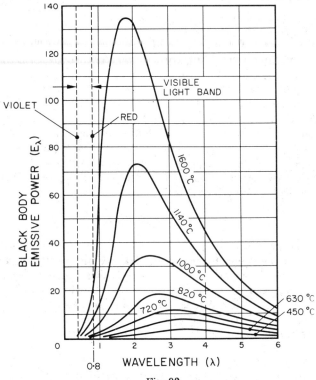

Fig. 92

The maximum emissive power corresponding to each wavelength is characteristic of a black body, but many materials are selective emitters, radiating relatively more strongly at some wavelengths than at others. It is the total radiation of thermal energy over all

wavelengths $\int_{\lambda_1}^{\lambda_2} E_\lambda d\lambda$ however that is important in heat transmission

problems, and the total emissivity ϵ is defined as the ratio of the total thermal radiation of unit area of the body over all wavelengths to the total thermal radiation of the black body at the same temperature over all wavelengths.

251

$$\int_{\lambda_1}^{\lambda_2} E_\lambda \, d\lambda$$

i.e. emissivity $\epsilon = \dfrac{\displaystyle\int_{\lambda_1}^{\lambda_2} E_\lambda \, d\lambda}{\displaystyle\int_{\lambda_1}^{\lambda_2} E_\lambda^{\,1} \, d\lambda}$

where $E_\lambda^{\,1}$ = emissive power of a black body

When radiated energy is received by a body it may reflect a part of it R, absorb some of it A, and allow some of it to pass through and be transmitted T. Thus, $R + A + T = 1$. Most engineering materials are opaque to thermal radiation so that $T = 0$ and we may write $R + A = 1$. Thus the ability of a material to absorb radiant energy depends upon the reflectivity of its surface, so that good reflectors are poor absorbers. KIRCHOFF'S LAW states that, for a particular wavelength, the ratio of the emissive power E_λ of a material to its absorptivity A_λ is unity

i.e. $\dfrac{E_\lambda}{A_\lambda} = 1$ or $E_\lambda = A_\lambda$

Thus a good reflector is a poor absorber and radiator of thermal energy, and a poor reflector is a good absorber and radiator. Hence lamp black, which has an emissivity of 0·97 for a range of temperature of $-18°C$ to $1650°C$ reflects very little of any radiant energy received, whereas polished steel, which has an emissivity ranging from 0·06 at $-18°C$ to 0·26 at $1650°C$, will reflect a high proportion of radiant energy received.

Inverse Square Law

For a body to receive radiant energy direct from a source it must 'see' the source, e.g. when warming one's hands beside the fire the radiant heat is cut off if an opaque surface is placed between the hands and the fire. The inverse square law states that the intensity of radiation received at a surface is inversely proportional to its distance from the source of radiation.

Consider a concave spherical surface S receiving energy E per unit time from a point source P. If the distance of the surface from the source is r, then the energy received per unit time per unit area of the surface S is given by $\dfrac{E}{S}$, where surface area $S = \dfrac{\theta}{2\pi} \cdot 4\pi r^2$

i.e. $S = 2\theta r^2$ where θ = solid angle of the cone

Therefore:

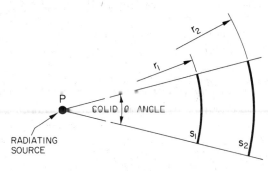

Fig. 93

$$\frac{\text{Intensity of radiation received at } S_1 \text{ at radius } r_1}{\text{Intensity of radiation received at } S_2 \text{ at radius } r_2} = \frac{\dfrac{E}{2\theta r_1^2}}{\dfrac{E}{2\theta r_2^2}}$$

$$= \frac{r_2^2}{r_1^2}$$

Stefan-Boltzmann Law

The work of Stefan in 1879, later established theoretically by Boltzmann, gave the law: *The total radiation emitted in unit time from a black body is proportional to the fourth power of its absolute temperature*:

i.e. $\dot{Q} = \sigma A T^4$

It has been stated that many bodies are selective transmitters, radiating relatively more strongly at some wavelengths than others. It is usual to refer to a 'grey' body which radiates at all wavelengths at a constant proportion of the radiation from a black body, so that it has the same emissivity ϵ at all temperatures. For a grey body

$$\dot{Q} = \sigma \epsilon A T^4$$
$$= 5 \cdot 67 \times 10^{-8} \epsilon A T^4 \text{ W}$$

Consider a grey body at temperature T_1 radiating in black surroundings which are at temperature T_2. Since all bodies are continuously receiving and radiating thermal energy:

Energy radiated from unit area of grey body to black surroundings per unit area. (All this energy is absorbed by the black surroundings) $= \sigma \epsilon T_1^4$
Energy radiated from black surroundings to

253

grey body and absorbed by the grey body $= \sigma T_2{}^4 \times$ Absorptivity

$$= \sigma \epsilon T_2{}^4$$

since for *all temperatures* emissivity $=$ absorptivity.

$\therefore$ Net radiation loss by $\dot{Q} = \sigma\epsilon(T_1{}^4 - T_2{}^4)$ per unit area grey body (if $T_1 > T_2$)

i.e. **Heat transmitted by $\dot{Q} = 5\cdot67 \times 10^{-8}\epsilon \ A(T_1{}^4 - T_2{}^4)$W radiation from grey body**

Provided the grey body is receiving very little radiation from its surroundings either because it is very small or because T_2 is relatively small, this expression may equally be used when the surroundings are not black.

No mention has been made here of many other aspects of radiation heat transfer, including radiation in gases, for which reference should be made to specialized works.

REVISION EXERCISES—CHAPTER 10

1. The inside of a cold room is maintained at a temperature of $-2°C$ when the surface temperature of its outer walls is $21°C$. The room is constructed of a 110 mm brick wall insulated externally by 75 mm of cork slabbing which is protected externally by wood 25 mm thick. Taking the thermal conductivities of brick, cork and wood as 0·95, 0·043 and 0·17 W/m degC respectively, calculate the leakage of heat per day of 24 hours through a 1 m square wall of the room. Estimate the temperatures at the interfaces.

   [Ans. 990 kJ; 19·3°C, -0·67°C)

2. Wet steam at $200°C$ is carried along a 50 mm diameter pipe which is lagged to a diameter of 100 mm. Taking the coefficient of thermal conductivity of the lagging as 0·17 W/m degC, estimate the heat loss per hour from a pipe 60 m long when the outer surface temperature of the lagging is $25°C$.

   (Ans. 58·25 MJ)

3. A 40 mm diameter pipe is insulated with 50 mm thick corrugated asbestos lagging for which the coefficient of thermal conductivity may be taken as 0·1 W/m degC. The pipe carries a liquid at $150°C$ and the outer surface temperature of the lagging is $37°C$. Estimate the heat loss per hour per 100 m length of pipe.

   (Ans. 20·4 MJ)

4. Obtain from first principles an expression for the heat flow through a thick cylinder in terms of the inner and outer radii,

the corresponding wall temperature and the coefficient of conductivity of the material of the cylinder.

A building is heated by circulating dry saturated steam at 110°C through 20 radiators. 400 m of 70 mm diameter pipe with 100 mm thickness of lagging connects the system and the outer temperature of the lagging is at 25°C throughout. If the flow of dry saturated steam leaving the boiler is 1000 kg/h and the heat loss from each radiator is 20 MJ/h, estimate the dryness fraction of the steam on return to the boiler. Take k for the lagging as 0·4 W/m degC and the latent heat of the steam as 2·2 MJ/kg.

(Ans. 0·715)

5. A furnace wall is made up of a 6 mm steel plate lined with 200 mm of refractory brick. The mean temperature of the brick surface inside the furnace is 1100°C. The coefficients of thermal conductivity of steel and brick are 40 and 1 W/m degC and the heat loss from the furnace through the wall is 7 MJ/m² per hour. Estimate the temperature of the steel surface and also the temperature at the interface.

(Ans. 711·1°C; 710·8°C)

6. A 1 m diameter spherical vessel contains a liquified gas at a temperature of −180°C. The vessel is lagged with two jackets each 100 mm thick, the coefficient of conductivity of the inner jacket being 0·05 W/m degC and of the outer jacket being 0·06 W/m degC. Find the quantity of heat leading into the vessel per hour when the outermost surface of the lagging is at a temperature of 15°C, and estimate the temperature at the joint between the lagging surfaces.

(Ans. 829·6 kJ; −57·76°C)

7. Working from first principles, obtain an expression for the rate of heat flow through the walls of a long thick cylinder, in terms of the inner and outer radii and the surface temperatures of the cylinder.

Calculate the heat lost per hour from the surface of a pipe, 30 m in length, carrying lagging of inner and outer diameters 80 mm and 125 mm respectively, when conveying wet steam at a temperature of 120°C. The external surface temperature of the lagging is 55°C and the coefficient of thermal conductivity for the lagging is 0·15 W/m degC.

The temperature drop through the wall of the pipe may be assumed to be negligible.

(Ans. 14·81 MJ)
[U.E.I.]

8. A building has a wall 225 mm thick of bricks with thermal conductivity 0·7 W/m degC. The inner and outer surfaces are at temperatures 21°C and 0°C respectively; what is the heat loss per square metre of surface.

   If a layer of wall board 6 mm thick with thermal conductivity 0·05 W/m degC is added to the inside and a layer of concrete 25 mm thick with conductivity 0·76 W/m degC is added to the outside, and the temperatures of the inside and outside surfaces of the building remain as before, what is the reduction in heat loss and the temperature of the inner face of the bricks?

   (Ans. 65·4 W/m²; 44·3 W; 15·7°C)
   [I.Mech.E.]

9. Distinguish between the terms emissive power and emissivity.

   A metal pipe of internal diameter 100 mm and thickness 5 mm carries steam at 120°C. If the pipe loses heat from its exterior surface at the rate of 40 W/m² degC excess temperature, find the mass of steam condensed in 1 m of pipe in 1 hour when the air temperature is 20°C and assuming that the outer surface of the pipe is at steam temperature.

   (Ans. 2·25 kg)
   [I.Mech.E.]

10. What is black-body radiation and how can it be obtained experimentally? State Stefan's law and define the term *emissive power* (or *emissivity*). Describe briefly some instrument suitable for detecting heat radiation and explain how you would use it to compare the emissive powers of different surfaces.

    Calculate the heat energy radiated per square centimetre per second from the surface of a tungsten wire at a temperature of 2187°C, assuming the emissive power of the tungsten is 0·35. Stefan's constant $\sigma = 5·67 \times 10^{-5}$ in C.G.S. degC units.

    (Ans. 17·4 cal cm⁻²s⁻¹)
    [I.Mech.E.]

11. State the fundamental law of heat conduction.

    A furnace wall is constructed of a layer of fire-brick 230 mm thick in contact with a layer of heat insulating material 115 mm thick. If the exposed surfaces of the fire-bricks and the insulating material are maintained at 1100°C and 90°C, respectively, calculate (a) the temperature of the surface of contact between the materials; (b) the heat flowing per second through one square metre of the wall.

    Assume k for fire-brick $= 1·25$ W/m degC and k for the insulating material $= 0·1$ W/m degC.

    (Ans. 961°C; 757·5 J)
    [I.Mech.E.]

INDEX

ABSOLUTE PRESSURE, 7
— temperature, 5
Adiabatic expansion: definition, 74
— —, work done, 77
Analysis of fuels, 129
— of products of combustion, 148
Anthracite, 129
Atomic weight, 16, 142
Avogadro's Hypothesis, 26
Axial thrust, 236

BITUMINOUS COAL, 129
Blade angles, impulse turbine, 233
Blading efficiency, 235
Boiler, efficiency, 196
—, Lancashire, 196
—, losses, 198
—, water-tube, 195
Bomb calorimeter, 130
Boyle's Law, 22
Boys' Calorimeter, 138
Brake power, 13, 50
—, thermal efficiency, 15

CALORIFIC VALUE, 13, 130, 135, 137
Calorimeter, bomb, 130
—, Boys', 138
—, separating and throttling, 182
Capacity, 13
Carbon, combustion of, 142
Carburation, 43
Carnot cycle, 100
Cetane number, 136
Characteristic gas equation, 24
Charles' Law, 22
Clearance volume, 12
Closed system, 2
Combustion, 141
Compression ratio, 12
Condenser, jet, 202
—, surface, 202

Conduction, 20, 243
Conservation of energy, 16
Constant volume cycle, 104
Convection, 20, 247
—, forced, 247
—, natural, 247
Cracking distillation, 135
Critical point, 177
Critical pressure, 231
Cycle, Carnot, 100
—, diesel, high-speed, 47, 111
—, diesel, slow-speed, 47, 109
—, four-stroke, 36
—, of operations, 10, 98
—, Otto, 104
—, two-stroke, 38

DALTON'S LAWS, 204
Delay period, 48, 111
Detonation, 108
Diagram, efficiency, 236
— factor, 219
Diesel cycle, 8, 109, 111
Dryness fraction, 170
Dual combustion cycle, 112

ECONOMIZER, 201
Efficiency, air standard, 105
—, boiler, 196
—, mechanical, 13
—, relative, 112
—, thermal, 14, 223
Emissive power, 249
Energy, balance account, 62, 223
— carried away by products of combustion, 147
—, conservation of, 16
—, general equation, 85
—, internal, 29, 176
—, kinetic, 1
— loss to chimney gases, 147

Energy (*cont.*), potential, 1
—, stored, 3
—, transitional, 3
Engine mechanism, 12
Enthalpy, 86, 169
Equivalent evaporation, 199
Exhaust gas, analysis, 148
— —, heat loss to, 198
Expansion, adiabatic, 74, 77
—, isothermal, 73
—, hyperbolic, 178
— of gases, 72
—, polytropic, 74
—, work done, 75

FLASH POINT, 135
Flow processes, 9, 84
— work, 84
Flue gases, analysis, 148
— —, energy, loss to, 147
Four-stroke cycle, 36
Friction power, 13
Fuel consumption, 13
Fuels, 128
—, calorific value of, 140

GAS CONSTANT, 24
— laws, 22
—, specific heats of, 10
Gauge pressure, 7
General energy equation, 85
Governing, cut-off, 222
—, throttle, 222

HEAT
—, mechanical equivalent of, 4
—, specific, 10
—, transmission of, 20, 243
—, unit of, 4
Heenan & Froude dynamometer, 53
Higher calorific value, by calculation, 156
— — —, by experiment, 137
Hydrogen, combustion of, 144
Hyperbolic expansion, 178
Hypothetical diagram, 216

IGNITION, PETROL ENGINES, 45

Impulse turbine, 228
Indicated power, 4, 55, 222
Indicator diagram, 42
Injection system, 48
Internal combustion engines, 35
Internal energy of gas, 29
— — of steam, 176
Isothermal expansion, 172

JET CONDENSER, 202
Joule's Law, 30

LANCASHIRE BOILER, 196
Latent heat, 169
Lignites, 129
Lower calorific value, 132

MASS FLOW, 229
Mean effective pressure, 57
Mechanical equivalent of heat, 4
Mole, 28
Molecular weight, 6, 142
Morse test, 59

NON-FLOW PROCESSES, 9, 75, 128
Normal temperature and pressure, 9
Nozzles, 228

OCTANE NUMBER, 136
Open system, 3
Orsat apparatus, 149
Otto cycle, 104

PARTIAL PRESSURE, LAW OF, 204
Petrol engines, 42
Petroleum, 134
Polytropic expansion, 74
Potential energy, 1
Power, brake, 13, 50
—, friction, 13
—, indicated, 13, 55, 222
—, pumping, 58
Pressure, absolute, 7
— gauge, 6
Processes—flow, 9, 84
—, non-flow, 9, 174, 178
Producer gas, 137
Products of combustion, 145

INDEX

Properties—of steam, 166
—, of systems, 4
Proximate analysis, 130
Pumping power, 58

RADIATION, 20, 250
Reaction turbine, 227
Relative efficiency, 112
Reversibility, 107
Rope brake, 51

SATURATED STEAM, 168
Separating and throttling calorimeter, 183
Slow-speed diesels, 47, 109
Specific fuel consumption, 13
Specific heat, of gases, 10
— —, ratio of, 31
— —, volume, 9
Standard temperature and pressure, 9
Steam, dryness fraction, 170
— engine, 215
— expansion, 178
—, internal energy of, 176
— plant, 193
—, properties of, 166
—, saturated, 168
—, superheated, 168
— tables, 168
— turbine impulse, 221
— turbine, impulse-reaction, 230

Steam (cont.)—volume, 170
—, wet, 168
Sulphur, combustion of, 147
Superheated steam, 168
Surface condenser, 202
Swept volume, 12
Systems, 2

TEMPERATURE SCALES, 5
Theoretical air supply, 144
Thermodynamics, first law of, 16
—, second law of, 101
Throttling, 180
— calorimeter, 182
Town gas, 137
Turbine, impulse, 227
—, reaction, 227
—, steam, 227
— velocity diagram, 234
Two-stroke cycle, 38

ULTIMATE ANALYSIS, 130
Universal gas constant, 27

VALVE-TIMING DIAGRAM, 38
Volume of steam, 170

WEIGHT, ATOMIC, 16
—, molecular, 16, 142
Willan's Line, 221
Work done by gases, 75